CAR BODYWORK IN
GLASS REINFORCED PLASTICS

Car Bodywork
in
Glass Reinforced Plastics

Richard Wood

PENTECH PRESS
London : Plymouth

First published 1980
by Pentech Press Limited
Estover Road, Plymouth
Devon

British Library Cataloguing in Publication Data

Wood, Richard
    Car bodywork in glass reinforced plastics.
    1. Automobiles – Bodies
    2. Plastics in automobiles
    3. Glass reinforced plastics
    I. Title
    629.2'6   TL255

    ISBN 0-7273-0304-X

Filmset by Mid-County Press, London SW15
Printed in Great Britain by Biddles of Guildford

Preface

Since the introduction of glass reinforced plastics in the early 1950s, the materials have found wide application for automobile bodywork construction both in the private and commercial sectors of the industry. Early users of glass reinforced plastics were attracted largely by the mouldability and good strength to weight ratio of the combination and the promise of design freedom by the elimination of expensive panel beating operations in the building of custom and short run work. An additional attraction was the comparative simplicity of the equipment required.

During the intervening years manufacturers of resins and reinforcements have solved many of the early problems associated with the application of the materials while experience on the part of users in various sectors has dispelled much of the euphoria generated by some of the early publicity given to their application.

A great deal of information exists in the literature regarding GRP laminates but it is scattered and because of the now very wide usage of the materials in other industries much of it is not applicable to the subject of car bodywork. This volume is believed to be the first to assemble the relevant information dealing specifically with the application of glass reinforced plastics to car bodywork and is intended to introduce the reader to the various methods of moulding a combination of glass fibre and polyester resin and to serve as a practical guide to the design of bodywork for the private vehicle.

Early chapters trace the introduction of the materials and list the many systems of moulding that have been developed over the past fifteen years from the initial, simple conception of hand lay-up to the now highly sophisticated resin injection systems devised for quantity production. The use of GRP for body components, hardtops, bodywork repairs and the problems of finishing plastics bodies is described. With the sophistication of moulding techniques much development work has been carried out by equipment manufacturers and a section is devoted to methods of mechanisation for production of both private and commercial bodies.

Later chapters describe the various forms of glass reinforcement available, the nature and handling of polyester resins and the design considerations specific to GRP work. The hand lay-up system of moulding is discussed in some detail for the benefit of the amateur constructor while other methods now used by some of the country's specialist constructors are detailed. The chapter on current work describes some of the more interesting work in progress in both the UK and USA, leading to a discussion on future trends and the specifics of quantity production of interest to the automobile student.

As a guide to more detailed information on the various aspects of the subject and to facilitate the task of the reader wishing to enter the field the book concludes with references to further reading and a list of suppliers of materials, equipment and ancilliaries.

The author would like to acknowledge assistance and information from Mr David Pamington, Editor 'Reinforced Plastics' and Mr Andrew Tait, Press Officer, Bayer UK Ltd, without which this book would have been less complete. My thanks also go to the many friends in the industry who have contributed information and in particular to Margaret Constable for her untiring help in typing the manuscript.

Richard Wood

Contents

1

GRP in automobile bodywork

1.1 INTRODUCTION

Glass reinforced plastics now generally termed GRP in the U.K. and FRP or fiber reinforced plastics in the U.S.A., are the most widely used group of materials in which glass fibre acts as a reinforcement to give the combination its strength. The combination has advantages as a structural material over both unreinforced plastics and glass fibres used by themselves. Unreinforced plastics have a low density, are relatively easy to process, are resistant to weathering and do not require a surface finish. However, factors which greatly limit their application as structural materials are their low stiffness, low strength and their tendency to 'creep' under load conditions. On the other hand glass fibres are very stiff and strong but are so brittle that these properties cannot be usefully utilised commercially. By combining glass fibres with plastics it is possible to obtain the advantageous properties of both in a single material — GRP. The result of the combination is a material which is light, strong with a weather resistant surface and which by the nature of its constituents can be fabricated into large and, if required, complex shapes. In GRP the glass fibres improve the otherwise low stiffness of the plastics, increase its strength and reduce 'creep'. The resin transmits load into the fibres and the whole being a matrix, the resin protects them from damage. Depending on the selection of the matrix material and the method of reinforcement the processor can produce a range of materials to suit the application.

There are two main types of plastics, thermoplastics and thermosets. The former type, such as polyethylene, polystyrene, nylon, etc., soften on heating and return to the solid state on cooling. Thermosets, which include epoxies and the polyesters are usually viscous liquids which set hard when activated by a catalyst and do not soften again on heating. Thus, a unique feature of GRP is that the final material can virtually be formed during the moulding process.

1

Another important reason for the rapid and wide acceptance of GRP particularly for large structures such as car bodies is that polyester and similar resins can be moulded without heat or pressure to effect cure and during the curing process they do not give off volatiles. This is because, unlike the majority of thermosetting plastics that cure or harden by a continuation of the chemical process by which they are produced, polyester and epoxides cure by a crosslinking reaction initiated by the catalyst which produces a slight exothermic heat but no volatiles.

Conventional thermosetting plastics require heat for the curing process during which gases and moisture are given off. For this reason they have to be moulded under conditions of both heat and pressure which in turn demands the use of heavy and expensive steel moulds and hydraulic presses capable of exerting many tons pressure. Polyesters and epoxides being viscous fluids set, or cure, simply by the addition of a crosslinking agent, or catalyst, which is mixed into the resin prior to its use. Thus it is not necessary to use high moulding pressures and, indeed, hand lay-up, or contact moulding can be carried out without the use of pressure during the lay-up or curing period. This characteristic has led to the development of a number of different methods of moulding which could not be used with conventional polymers.

In recent years the development of pre-impregnated glass fibre/-polyester moulding materials (SMC) has encouraged the larger scale manufacturers to use GRP materials to an increasing extent for components such as bumpers and grilles.

The various types of polyester resin and their application in conjunction with different reinforcements discussed in later chapters will be better understood by the brief description in Chapter 2 of the manner in which polyesters are produced.

1.2   BASIC PROPERTIES

In GRP the combination of a comparatively brittle resin and high tensile strength glass fibres produces a laminate or moulding which has excellent properties from the point of view of tensile, shear and compressive strength. GRP laminates or mouldings also exhibit good resistance to denting and if components are suitably designed do not warp even under cyclic conditions of temperature and humidity. Additionally, if correct combinations of resin and glass type are selected and correctly moulded they do not suffer weathering effects and thus do not require protection against corrosion. A further property and one of major importance to the automobile body designer, is that GRP laminates are extremely light as compared with steel panels of equal strength.

Although it can sometimes be a combination of non-mechanical properties which lead to the use of GRP in a particular project, for example, ease of fabrication, it is invariably the strength of the material which confers success on the application. Indeed, as compared with a conventional body construction material such as mild steel, a GRP laminate can have a strength/weight ratio factor 2–5 times higher but a stiffness/weight ratio factor some 2–3 times lower. This high strength/low stiffness characteristic of a GRP laminate means that in a body deformation and buckling can occur. Structural design is thus influenced more by the stiffness than the strength factor.

This property of the material, as shown in later chapters, can be overcome by designing shape into the mouldings and by avoiding large, generally flat areas. With this material when failure does occur it is by brittle fracture thus the major concern of the designer should be in the area of stiffness.

1.3   DESIGN FREEDOM

Although resin/glass fiber laminates are more expensive than traditional materials due to the cost of producing the components of fibre and resin, this factor can usually be offset by the greater design freedom offered and by the comparative simplicity with which complex shapes can be moulded. To the body designer the fact that the raw material is virtually in the fluid state permits greater freedom to produce difficult contours and complex double curvature without recourse to the use of costly, time consuming and highly skilled operations of wheeling and panel beating. This freedom of shape — an important factor in body design where fashion dictates the use of flowing double curvature panels — is coupled in a GRP moulding with the facility to produce large and complex structures in one piece. This factor is again of particular importance from an economic point of view as it can eliminate labour intensive assembly operations and the need for costly jigs and welding fixtures associated with conventional body construction in metal.

However, although characterised by a high strength to weight ratio, GRP laminates, it should be understood, do not have the same flexural rigidity as steel or aluminium. Thus successful application is very dependent upon a full knowledge of their properties by the designer. The major properties of polyester resins with various forms of glass fibre reinforcement as compared with metal are given in Table 1.1.

Table 1.1 COMPARISONS OF GENERAL PHYSICAL AND MECHANICAL PROPERTIES OF GRP MOULDINGS PRODUCED BY DIFFERENT METHODS WITH METALS (COURTESY BRITISH INDUSTRIAL PLASTICS LTD.)

Material and method of production	Material cost per kg relative to hand lay-up mat (mat)	Density (kg/m³)	Flexural modulus (GPa)	Tensile modulus (GPa)	Flexural strength (MPa)	Tensile strength (MPa)	Impact strength Izod (notched) (J/m)	Thermal conductivity (W/mK)
SMC (hot press moulding)	1.43	1854.5	10.34	11.03	172.37	69.00	373.0	0.1586
Chopped strand mat (matched die moulding)	1.09	1522.4	10.69	10.34	241.32	124.11	692.0	0.1586
Continuous filament mat (cold press moulding)	1.03	1439.4	6.9	8.27	172.37	89.6	480.0	0.1730
Hand lay-up (mat)	1.00	1494.7	8.27	8.96	194.00	103.40	640.0	0.2020
Spray-up (roving)	0.89	1494.7	8.27	8.96	194.00	103.40	640.0	0.2020
Carbon fibre	–	1540.0	–	193.0	517.00	724.00	2047.0	14.4
Mild steel	0.23	7805.8	207.55	200.0	193.95	206.84*	1066.0	46.15
Stainless steel	2.46	7896.4	193.95	193.95	220.66	220.06*	1066.0	128.3
Aluminum	2.06	2712.6	69.00	69.00	82.7	82.4†	320.0	191.83

* Yield stress
† 0.1% proof stress

Note:
SMC at 25% glass content
Matched die moulding at 40% glass content
Continuous filament mat cold press moulding at 20% glass content
Hand lay-up at 30% glass content
Spray-up at 28% glass content
40% carbon fibre reinforcement
Average values taken for physical properties.

While the properties of high strength to weight ratio and freedom of shape are of immense value to the body designer, another important advantage which is not available when working in sheet metal is the facility to vary at will the thickness of the material where required. Although this is more difficult to achieve when using some of the moulding methods available, as will be discussed later (Chapter 4) with reinforcement in the form of woven cloth, rovings, or the more widely used chopped strand mat, it can quite readily be achieved by the addition of extra layers of reinforcement. In this way, areas where stresses are high or additional stiffness is required, for example attachment points, thickening of the laminate can be carried out locally without incurring a weight penalty over the general structure. Another feature of GRP laminates are the advantages they offer in service. Apart from freedom from corrosion — a factor at the top of the list of causes of deterioration in steel bodies — a moulded body can be repaired comparatively simply without the need for the speciallist welding and panel beating skills required for work on its metal counterpart.

1.4 RELATIVE COSTS

In comparing the cost of a GRP body with its traditional steel counterpart consideration must be given to the scale of production. In a direct comparison with a conventional mass produced steel body the moulded component will be considerably more expensive due not only to the higher cost of the materials but also the fact that the moulding process is more time consuming and, in general, more labour intensive. For these reasons, although the price of both resins and glass reinforcement is considerably less than when first introduced in the 1950s, the moulded GRP body has never displaced the conventional pressed steel body in large scale production.

When shorter runs of more specialised bodies are considered, GRP comes into its own. Firstly tooling costs are very much lower and, secondly, few concessions to form have to be made whereas steel body tool design places many limitations on shape and often requires many separate sections to be produced for assembly by welding or rivetting. Thirdly, because of the nature of the tooling, a moulded body can be put into production in considerably less time than a pressed steel or aluminium structure.

Thus the use of GRP has been accepted by companies working in the more specialised area of body construction where a large investment in tooling would not be economic and could not be amortised over the limited number of bodies of any one type produced. An idea of the relative cost GRP and commonly used metals is given in Table 1.2.

1.5 DEVELOPMENT OF GRP

Since the introduction of reinforced plastics, very considerable development has been carried out by both private and commercial bodybuilders. The early methods of hand lay-up have in many cases been superseded by more sophisticated systems of spray-up, vacuum bag, resin injection, matched and the pressure moulding of sheet moulding compounds (SMC). The methods and the material used are discussed in later chapters. The earliest method of moulding was the hand lay-up or contact process which owes much of its initial development to the aircraft industry where in the late 1940s it was first used for the production of randomes, wing tips, doors and similar large and relatively high strength components. Here the ability of the process to provide local thickening and stiffening where stresses are high and yet avoid a weight penalty where stresses are low was of immense value. Another factor which played a large part in the acceptance and development of the process for specialised automobile work was that short runs of highly complex parts could be produced with the minimum of tooling.

Unlike the automobile mass production industry where output of a popular car is measured in tens of thousands and economics favour the use of complex and often automated tooling to achieve production targets and to reduce man hours per body, the aircraft industry is of necessity highly labour intensive. Here components have to be produced to the very highest of standards and although the cost of a single aircraft is very high and requires hundreds of such components the total number of each is comparatively small. For these reasons a tooling system which could be built quickly and inexpensively, but had a sufficiently long working life to produce to the requisite tolerances the short runs of high strength to weight components involved, was obviously attractive.

For the majority of components moulds could be produced from wooden or plastic master patterns using glass reinforced polyester or epoxide resins — a method of construction which had the added advantage that modifications at the prototype stage could be incorporated rapidly and with the minimum of cost. Consumption of GRP in vehicle construction in W. Europe in 1976 is given in Table 1.3.

Table 1.2 COMPARATIVE THICKNESS, WEIGHT AND MATERIAL COST FOR EQUAL STRENGTH PROPERTIES (COURTESY SCOTT BADER LTD.)

Material	Cost		Equal tensile strength			Equal tensile stiffness			Equal bending stiffness		
	Per unit volume	Per unit weight	Thickness	Weight	Cost	Thickness	Weight	Cost	Thickness	Weight	Cost
Mild steel	1	1	1	1	1	1	1	1	1	1	1
Aluminum	2	6	1.8	0.6	4	3	1.1	6	1.5	0.5	3
Stainless steel	15	14	1	1	14	1.1	1.1	14	1.1	1.1	14
Random GRP	1	5	2.4	0.5	2.5	25	5	25	3	0.6	3
Unidirectional GRP	1.4	6	0.3	0.1	0.6	6.8	1.5	9	1.9	0.5	3

Table 1.3   CONSUMPTION OF GRP IN VEHICLE CONSTRUCTION IN
W. EUROPE IN 1976 (COURTESY GEVETEX TEXTILGLAS GmbH)

Country	Consumption (tons)
Austria	600
Benelux	1,700
Gt Britain & Ireland	15,600
Denmark	500
Finland	800
France	20,300
W. Germany (FGR)	13,500
Italy	7,200
Norway	400
Spain	2,700
Sweden	2,400
Switzerland	800

1.6   BASIC MOULDING METHODS

The various methods of moulding GRP components and bodies are
described in some detail in Chapter 4. The brief description given here
is intended only to indicate the scope of the processes.

1.6.1   Hand lay-up

Hand lay-up or contact moulding is the simplest method of
producing resin/glass components and lends itself particularly to
short run production of large mouldings such as car bodies, boat
hulls, caravan and commerical vehicle cabs. A single female mould is
used in the majority of cases as this method gives a finished outer
surface on the component (Figure 1.1). The mould itself can be of
GRP, built up of successive layers of reinforcement, usually chopped
strand mat over a suitably waxed master pattern. The laminate, also,
is composed of successive layers, most commonly of chopped strand
mat, each layer impregnated with catalysed and accelerated resin
which are rolled by hand using special rollers to consolidate the
reinforcement and to drive out any entrapped air. When the resin has
gelled, but not fully cured, the surplus laminate is trimmed off, to the
edges of the mould, the laminate allowed to cure fully after which it
can be freed and lifted from the mould.

Resin
Laminate
Roller
Mould
Gel coat

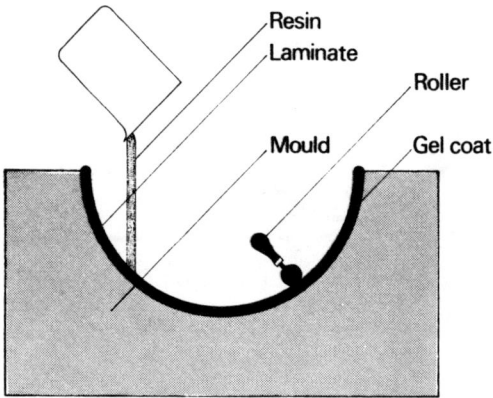

Fig. 1.1 *Hand or contact lay-up moulding*

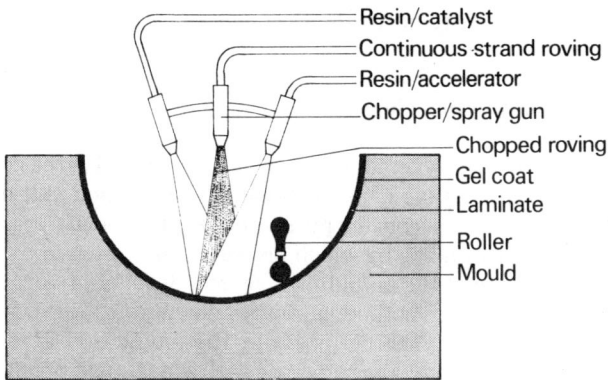

Resin/catalyst
Continuous strand roving
Resin/accelerator
Chopper/spray gun
Chopped roving
Gel coat
Laminate
Roller
Mould

Fig. 1.2 *Spray-up moulding*

1.6.2 Spray-up

As hand lay-up applications progressed constructors sought a faster system for applying the reinforcement and resin to a single mould. This led to the design of 'spray-up' guns which permit resin by itself for gel coats or both chopped reinforcement and catalysed and accelerated resin to be sprayed simultaneously into the mould until a layer of the required thickness is built up (Figure 1.2). While giving a very much higher rate of production than the hand lay-up method, the quality of the moulding is still dependent upon the skill of the operator as regards consistency of laminate thickness and as with

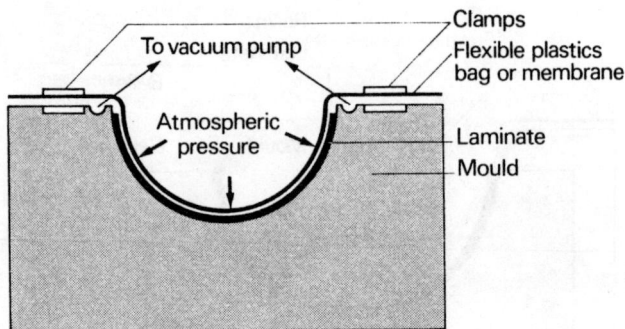

Fig. 1.3   Vacuum bag moulding

hand lay-up, consolidation of the resin/glass is effected by hand rolling. It will be appreciated that considerable skill is also required on the part of the operator to ensure the thickness of the laminate is consistent or that the requisite thickening is applied where necessary.

1.6.3   Vacuum bag moulding

The vacuum bag method of moulding which originated in the aircraft industry is an alternative to hand lay-up. This method still uses a single mould but differs from the purely contact method in the use of a membrane or vacuum bag by which pressure can be exerted over the entire inner surface of the moulding (Figure 1.3). After lay-up, either by contact or spray-up method, a plastics membrane is draped over the laminate, sealed at the periphery of the mould and a vacuum drawn from within pulling the bag down with a maximum pressure on the laminate of one atmosphere. The method gives a well consolidated laminate with a better glass-to-resin ratio than usually can be obtained by the contact process.

1.6.4   Pressure bag moulding

An alternative method which operates on a similar principle to the vacuum bag process is the pressure bag system. It uses a single mould and a flexible membrane to exert pressure on the inner surface of the moulding. Here the open surface of the mould is sealed with a rigid and air tight cover and air pressure is introduced through the cover plate to press the membrane on to the surface of the laminate (Figure 1.4). When this method is used with a GRP mould suitable for the

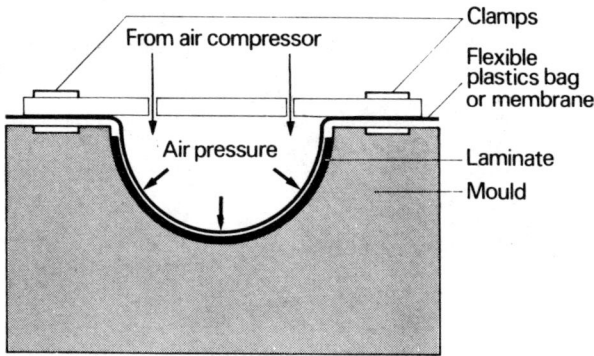

Fig. 1.4 *Pressure bag moulding*

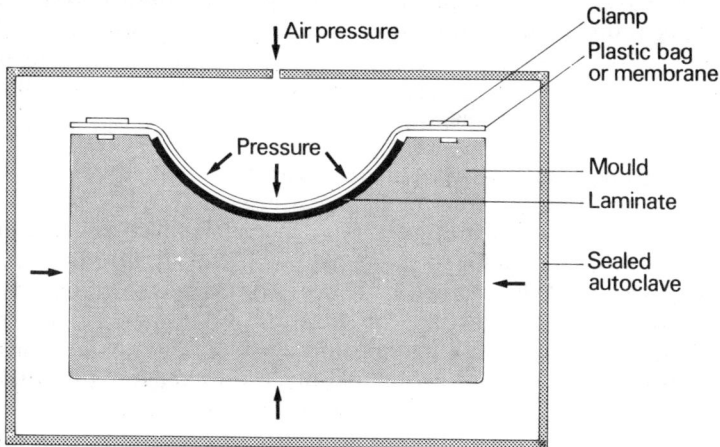

Fig. 1.5 *Autoclave moulding*

vacuum process the pressure applied must be limited and care must be taken to avoid mould distortion.

1.6.5 Autoclave moulding

This is a variant of the pressure bag method in that it uses air pressure to consolidate the laminate but is always used in conjunction with heat to speed the curing process. In the process the entire mould with its covering bag or membrane is placed in a sealed and heated chamber which is then pressurised to force the membrane against the laminate (Figure 1.5). The process gives a good surface on the inner

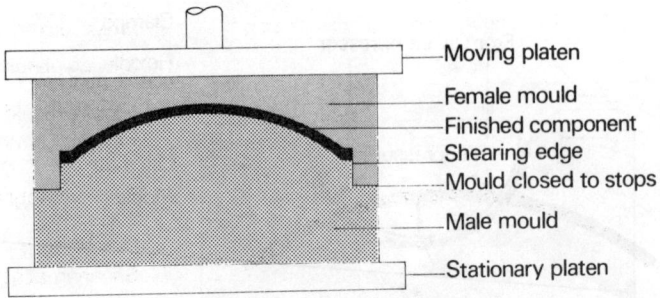

Fig. 1.6   *Cold press moulding*

side as do the other bag moulding methods but it entails considerably higher equipment cost. As the pressure of the air is constant around the mould this can be of lighter construction than when pressure is confined to the inner surface only.

1.6.6   Cold press moulding

The cold press method is sometimes used as an alternative to the hand lay-up process. Cold press moulding is essentially a low pressure process in which either strong GRP or light metal tooling can be used in conjunction with a low pressure hydraulic press (Figure 1.6). The process relies on the exothermic heat generated by the gelling and curing of the resin and has the advantage over hand lay-up of producing mouldings with smooth inner and outer surfaces. Costwise it falls between the hand lay-up and hot press moulding processes described later.

1.6.7   Hot press moulding

This process is now used where long production runs of medium sized components such as bumpers, grilles and panels are required. Tooling consists of heated matched metal moulds which are mounted in a low pressure hydraulic press. Reinforcement can take the form of continuous strand mat or a preform made in an earlier process. The moulds can be used either with the female half on the lower platen of the press or for more complex mouldings the preform is placed over the male which is then located on the lower platen (Figure 1.7). In operation, a measured quantity of catalysed resin is poured over the preform and the press closed until gelling takes place. The component is then removed and allowed to cure fully.

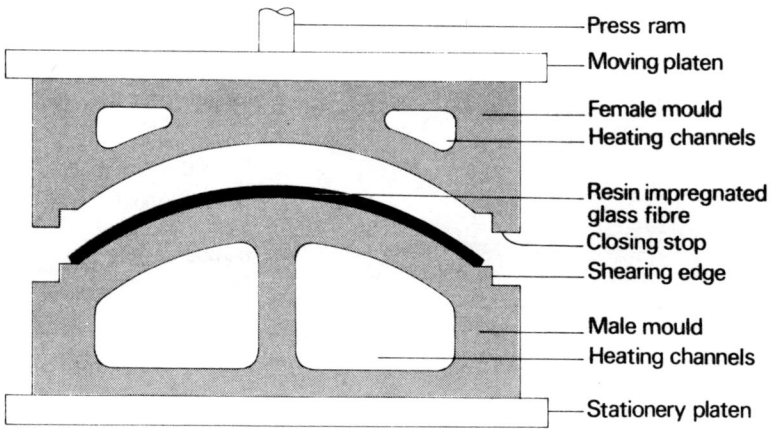

Press ram
Moving platen
Female mould
Heating channels
Resin impregnated glass fibre
Closing stop
Shearing edge
Male mould
Heating channels
Stationery platen

Fig. 1.7 *Hot press moulding*

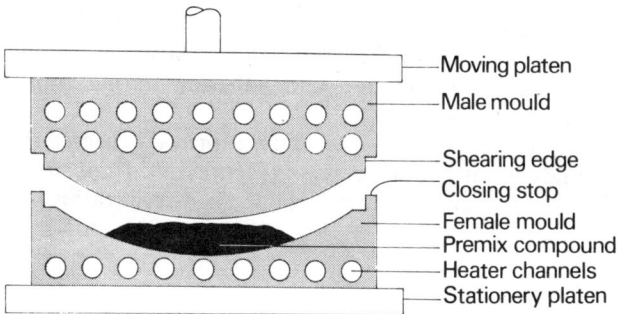

Moving platen
Male mould
Shearing edge
Closing stop
Female mould
Premix compound
Heater channels
Stationery platen

Fig. 1.8 *Dough moulding*

1.6.8 Dough moulding

Dough or premix moulding is used for producing smaller components in quantity. It has the advantages of speed and consistency of wall thickness and weight of the finished component. Glass reinforcement is mixed with resin pigment, filler and catalyst. The compound is placed in the mould cavity and moulded under both heat and pressure (Figure 1.8). Heated matched metal moulds are used in conjunction with an hydraulic press giving from 7–105 kgf/cm^2 pressure.

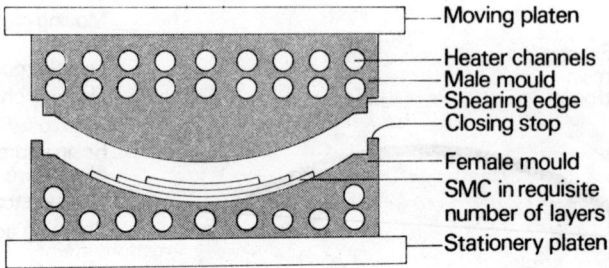

Fig. 1.9   SMC moulding

1.6.9   Sheet moulding compound (SMC moulding)

This technique has greatly increased in use for the moulding of automobile body components by large scale producers. Sometimes referred to as 'prepreg' moulding the material is supplied in the form of sheets of glass fibre reinforcement impregnated with a catalysed polyester resin and sandwiched between two layers of polyethylene film to prevent contamination and adhesion during transit and storage.

In use, sheets are tailored to size and are formed in heated matched metal moulds in an hydraulic press (Figure 1.9). High production speeds can be obtained with predictable glass/resin ratios.

1.6.10   Resin injection

A more sophisticated method of moulding large components is the injection of catalysed resin into a closed mould previously loaded with a carefully adjusted quantity of reinforcement. In some cases injection of the resin is assisted by the application of a low vacuum (Figure 1.10). Where this method is employed it is important that perfect sealing around the edges of the male and female moulds is obtained otherwise air will be drawn in which will be trapped in the moulding. Moulds for resin injection can be of GRP suitably stiffened but for medium production runs it is usual to employ metal moulds. In this process the two halves of the mould can be bolted or even clamped together thus obviating the need to use an hydraulic press.

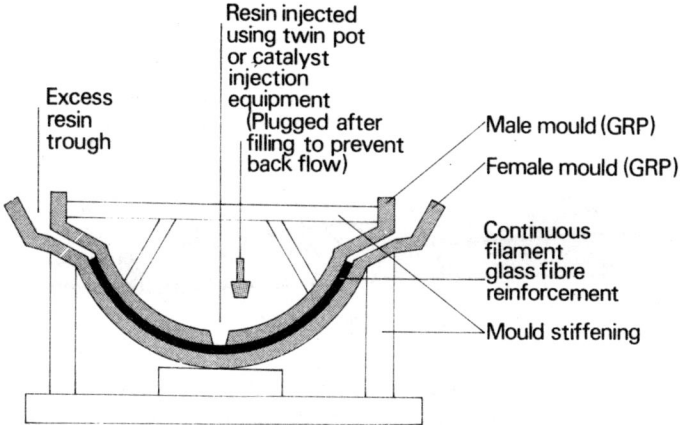

Fig. 1.10   *Resin injection moulding*

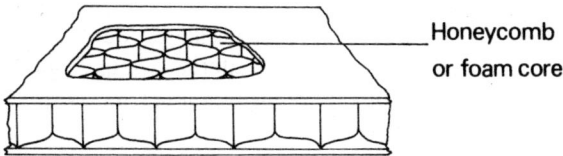

Fig. 1.11   *Composite moulding with honeycomb core*

1.6.11   Composite moulding

Composite structures incorporating resin/glass fibre reinforcement can be made in many ways, a number of which are described in greater detail in Chapter 4. One of the earliest methods, which originated in the aircraft industry in the production of radomes, is the use of a double skin of GRP separated by a phenolic resin impregnated paper honeycomb material. A less sophisticated version of a double skin structure uses a lightweight core of blocks of rigid foam or balsa wood. Later development of the system incorporates aluminium honeycomb in combination with epoxide/glass laminate skins, Figure 1.11. These methods are all carried out by means of hand or spray-up processes. Another method of producing a rigid structure with a lightweight core is by in situ foaming of a double skinned component using one of the foaming liquid polyurethane formulations. One of the latest processes known as the 'reservoir' process is the use of a soft open-cell foam as the carrier for the resin which when packed between

Polyethylene
film

Woven glass
tape or cloth

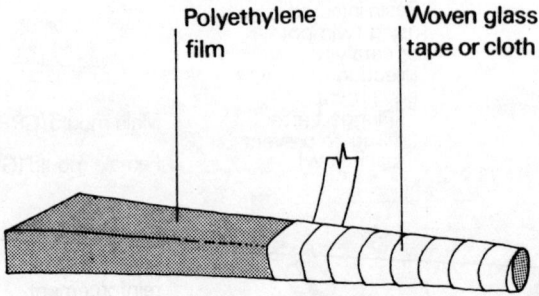

Fig. 1.12   *Duct winding*

two layers of dry glass fibre reinforcement in a closable matched mould is squeezed out forming a double skinned structure with a foam core.

1.6.12   Duct winding

The winding of woven glass cloth tape over collapsible or destructible plaster cores is used to produce complex hollow components such as heater ducts and similar items (Figure 1.12).

2

Materials: resins, catalysts, accelerators, pigments and fillers

2.1 POLYESTERS

Polyester resins are brought to a final state of cure by the addition of a catalyst which permits reaction of the styrene monomer contained in the resin with other ingredients to produce a molecular cross linked polymer. Full understanding of the reaction forces are of vital importance to the successful use of the various methods of moulding. In the past many failures have resulted from a lack of appreciation of the factors that produce gellation and final cure.

2.2 CATALYSTS

Catalysts most widely used are organic peroxides, methylethyl ketone peroxide or benzoyl peroxide. The latter is the most commonly used in hand lay-up work. The use of a peroxide catalyst alone usually demands the addition of heat to effect a reasonably rapid cure in most moulding applications. Where cure at shop temperature is required and this is usually the case in a body plant, an accelerator is added which has the effect of activating the catalyst at a lower temperature.

Accelerators for use with methylethyl ketone peroxide catalysts are cobalt derivatives and combinations based on them. Where benzoyl peroxide catalysts are used accelerators are based on amines such as dimethylaniline. Table 2.1 lists commonly used catalysts and accelerators that are used with them. It is essential to select the right type of catalyst and accelerator combination as well as the correct amount. Resin suppliers give recommendations which will ensure optimum properties in the finished laminate. Some examples for both cold and hot curing systems are given in Sections 2.3 and 2.4.

Table 2.1 COMMONLY USED CATALYST AND ACCELERATOR SYSTEMS. (COURTESY BRITISH INDUSTRIAL PLASTICS LTD.)

Catalyst	Accelerator or promoter	Normal operating temperatures	Moulding process	Remarks
Methylethylketone peroxide	Cobalt	15–25°C	Hand lay-up and catalyst injection	Promoter usually cobalt nephthenate or cobalt siccatol
Benzoyl peroxide	Amine	10–25°C	Mainly cold press and resin injection. Occasionally hand lay-up	Promoter generally an aromatic amine
Cyclohexanone peroxide	Cobalt plus amine	15–25°C	Body stopping	Paste form suitable for easy dispensing and mixing
Benzoyl peroxide	None	100–150°C	Matched metal moulding	Not highly active below 80°C

Table 2.2 EFFECT CATALYST, ACCELERATOR AND INHIBITOR ON GEL TIMES WITH REFERENCE TO A BENZOYLE-PEROXIDE-AMINE SYSTEM USING HYDROQUINONE AS INHIBITOR. (COURTESY BRITISH INDUSTRIAL PLASTICS)

	Approximate gel time at:	
	20° C	*100° C*
Polyester resin (65% in styrene) without inhibitor	2 week	30 min
Polyester resin (65% in styrene) containing added inhibitor (0.01% hydroquinone)	1 year	5 h
Polyester resin (65% in styrene) containing added inhibitor (0.01% hydroquinone) and catalyst (1% benzoyl peroxide)	1 week	5 min
Polyester resin (65% in styrene) containing added inhibitor (0.01% hydroquinone), catalyst 1% benzoyl peroxide) and accelerator (0.5% dimethyl aniline)	15 min	2 min

Most resin manufacturers supply polyesters which contain a proportion of accelerator. These resins havé a storage life of several months at shop temperature and are the most suitable for cold-cure applications requiring only the addition of the correct proportion of catalyst immediately prior to use. Table 2.2 shows the effect of catalyst, accelerator and inhibitor.

2.2.1 Curing reaction

The cure of a polyester commences as soon as the catalyst is added with the speed of cure dependent upon the type of resin and the activity of the catalyst. The reaction, as previously mentioned, is exothermic although in a laminate which has a comparatively large surface area, the temperature rise is not excessive and should not cause any problems.

The curing reaction actually takes place in three stages. Gel time is the period from the addition of the catalyst or accelerator depending upon which system is used, until the setting of the resin into a soft gel. Hardening time is the period from gellation to a point where the resin is sufficiently hard to allow the moulding to be removed from the mould. The third phase is the maturing time which, dependent upon the particular system used and the ambient temperature may be hours, days or in some cases even weeks. At this point the moulding will acquire full hardness, chemical resistance and dimensional stability. Figure 2.1 shows equivalent post curing times and equivalent temperatures.

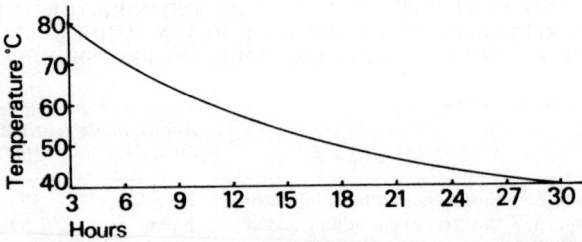

Fig. 2.1   Equivalent post curing times and equivalent temperatures. (Courtesy Scott Bader Ltd)

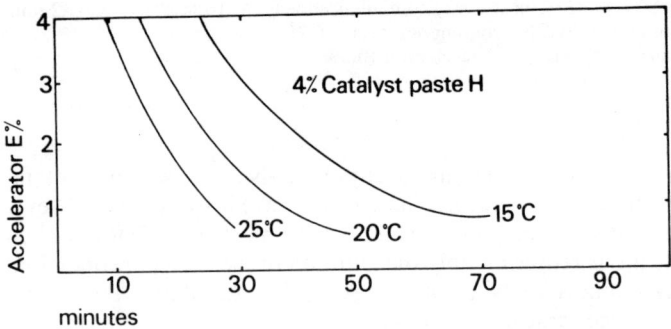

Fig. 2.2   The pot life of fast gelling Crystic resins. (Courtesy Scott Bader Ltd)

2.3   COLD CURING FORMULATIONS

This system is the most widely used for open moulds, by the amateur bodybuilder and in specialist shops where rapid mould turnround is not required. As previously mentioned, to obtain cure at shop temperature both a catalyst and an accelerator must be used. Crystic polyester resins, to cite one specific example, can be catalysed by Catalyst Paste H, Catalyst liquid H, Catalyst L, M or O. Catalyst Paste H is a stable dispersion of cyclohexanone peroxide which remains effective indefinitely. Catalysts L, M and O are liquid dispersions of various methylethylketone peroxides, termed MEKP which differ only in degree of activity and hardening rates. It should be noted that the pot life of Crystic resins containing Catalyst Paste H, without any accelerator have a usable pot life of some 8 hours at shop temperature. The pot life with liquid catalysts is some 2 hours shorter. Figures 2.2–2.5 show the pot life of various Crystic resins.

The accelerators used with resins catalysed with Paste or Liquid H or catalysts L, M or O is Accelerator E, G or R. The most widely used

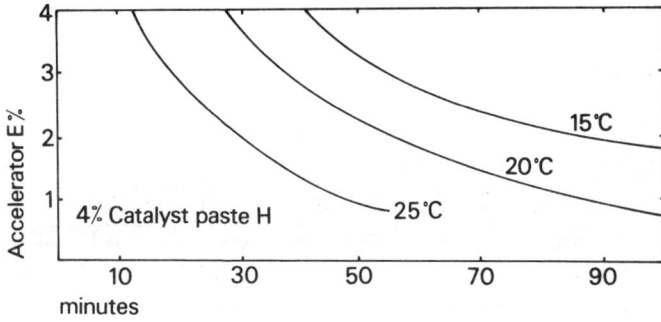

Fig. 2.3   *The pot life of medium gelling Crystic resins.* (*Courtest Scott Bader Co. Ltd*)

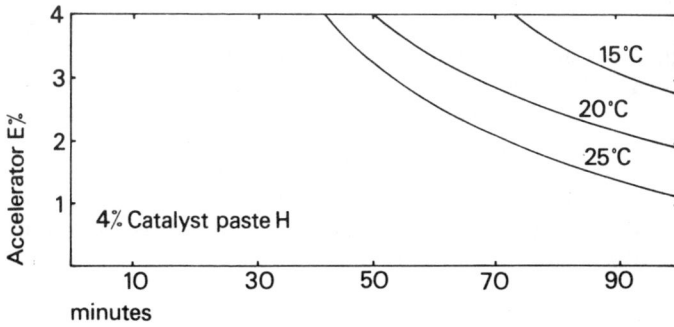

Fig. 2.4.   *The pot life of slow gelling Crystic resins.* (*Courtesy Scott Bader Co. Ltd*)

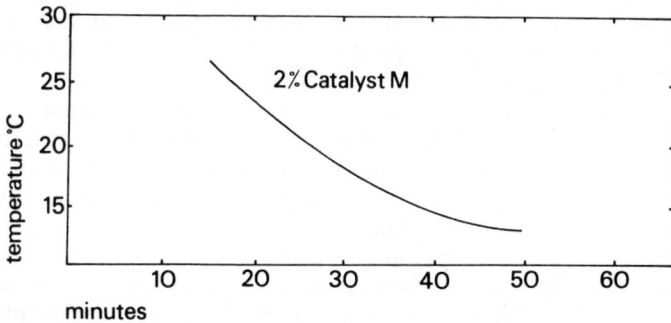

Fig. 2.5.   *The pot life of pre-accelerated Crystic resins.* (*Courtest Scott Bader Co. Ltd*)

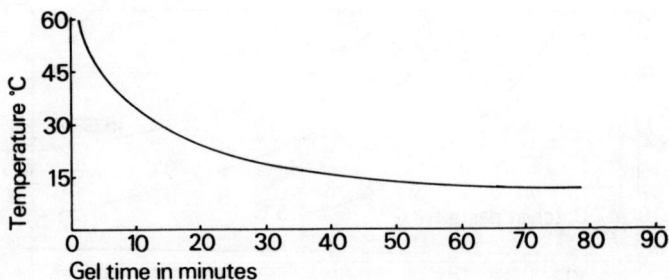

Fig. 2.6. The effect of ambient temperature on the gel time of a cold cure polyester.
(Courtesy Scott Bader Co. Ltd)

is Accelerator E which contains 0.4% of cobalt as the octoate in a styrene solution. Accelerators G and R contain 1% and 6% of cobalt octoate respectively.

In use, the quantity of Accelerator E is usually between 1–4% by weight of resin. The actual proportion depends on the gel time required and the shop temperature. The gel time is always controlled by varying the proportion of accelerator in the resin and *not* by altering the amount of catalyst used.

The effect of ambient temperature on the gelling time of a typical cold cure polyester resin is shown in Figure 2.6. The lowest temperature at which curing should be carried out is 15°C. Lower temperatures will tend to cause undercuring.

2.4 HOT CURING FORMULATIONS

The use of hot curing systems is invariably confined to production shops. They can be applied to complete body mouldings whether produced by hand or spray lay-up or by other methods detailed in Chapter 4, but are today more generally applied to matched moulding processes for the production of body panels and components where large runs are required.

For Crystic polyesters Catalyst Powder B which contain 50% benzoyl peroxide is recommended. In most formulations 2% by weight is added. This gives a pot life of about seven days at shop temperature. Laminates can be cured at 100–140°C for from 1–10 min. The actual period will depend on the thickness of the laminate, the specific type of resin used and on the heat capacity of the moulds. While overcure is not possible, the temperature should not be allowed to exceed 140°C. The effect of temperature on the setting time of a typical polyester is shown in Figure 2.7.

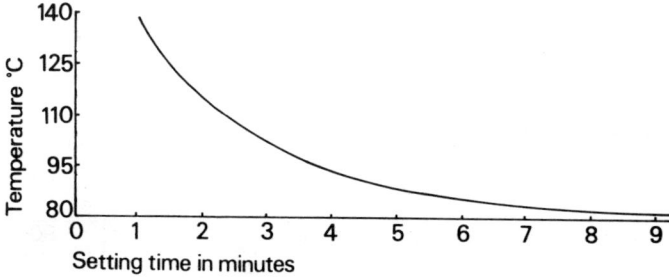

Fig. 2.7.   The effect of temperature on the setting time of a typical polyester. (Courtesy Scott Bader Co. Ltd)

2.5   GEL COAT RESINS

In automobile body building work the gel coat is an important area of the laminate both from the point of view of appearance and weathering. It is, however, the most vulnerable part and special care must be taken in the choice of resin and in the mixing of accelerator and of filler and/or pigments if these are to be used. Gel coats can be applied by brush, paint roller or spray — when spraying the gel coat a specially formulated resin is used in conjunction with catalyst injection type spraying equipment. One important factor in gel coat application is that the materials should be allowed to attain workshop temperature before being formulated.

Most resin manufacturers supply ready-coloured, pre-accelerated gel coat resins which enable the laminator to eliminate much of the labour-intensive mixing process together with the possibility of operator error when the addition of very small proportions of pigment and accelerator are involved.

Gel coat resins supplied by the resin manufacturer are formulated to give optimum characteristics including freedom from drainage on near vertical surfaces, adequate coverage with good wetting properties, rapid gelling and good adhesion to the main laminate. Thus any major modification to the resins by the fabricator will in all probability upset the balance of properties achieved by a critical combination of flow additives during manufacture. The recommended catalyst for gel coat resins is a 50% solution of a medium reactivity MEKP[1] which disperses easily in the resins. Low reactivity MEKP catalysts are not recommended for use with gel coat resins. This also applies to the reduced hazard catalysts because for some applications it is not possible to obtain optimum cured properties from gel coats cured with these preparations.

As recommended, gel coat application should not be carried out in

Table 2.3 POT LIFE OF 100 g OF NON-ACCELERATED CRYSTIC GEL COAT RESIN CONTAINING 2% CATALYST M AND 1–4% OF ACCELERATOR E (COURTESY SCOTT BADER LTD.)

Parts of Accelerator E to 100 parts catalysed. Crystic gel coat resin	15°C (min)	Pot life at 20°C (min)	25°C (min)
1	62	38	24
2	40	26	18
3	31	19	13
4	24	14	8

shop temperatures below 15°C. Above 18°C it is possible to use less catalyst than at the lower temperature. It should be noted however, that these conditions do not apply when using low fire hazard gel coats[1]. These resins, are now available in formulations suitable for automobile body moulding where fire resistance is important and can be obtained for brush or spray application. Some are available in a range of standard colours: others can be pigmented by the moulder if desired but it is stressed by manufacturers that care must be taken in selecting suitable pigments and it is advisable to test pigment pastes for colour stability in the resin prior to full scale use. A guide to pot life of these gel coat resins are given in Table 2.3. The actual gel time will depend on a number of factors including the material used to construct the mould. Typical properties of Crystic gel coat resins are given in Table 2.4.

2.6 PIGMENTS

Pigments are normally supplied in the form of resin pastes which are readily mixable in both gel coat and laminating resins if the former are supplied uncoloured. It should be noted that the addition of pigments offset the gel time of the resin. Some colours, notably black and most blues lengthen it considerably. Indeed too high a proportion of black can inhibit gelling totally. Other colours shorten gel time and thus the minimum amount of pigment should be used. Up to 10% by weight is recognised as a permissible maximum for most colours.

Another factor to be taken into account in the use of pigments is that some pigments are fugitive (ie, change colour or disappear), others contain a proportion of dye which may be sensitive to the catalyst used and can be oxidised causing bleaching. For these reasons when using pigments other than those recommended by the

Table 2.4 TYPICAL PROPERTIES OF CRYSTIC GEL COAT RESINS (COURTESY SCOTT BADER LTD.)

Crystic	Specific gravity at 25°C	Acid value (mgKOH/g)	Volatile content (%)	Appearance	Stability* in the dark at 20°C (months)	Gel time at 25°C (min)	Hardness Barcol	Water absorption (BS 2782: Part 5: 502G) (mg)
Gelcoat 1 PA	1.12	21	33	Brown opaque	3	9†	50	17
Gelcoat 33 PA	1.15	19	33	Cloudy	6	9†	38	20
Gelcoat 33	1.15	19	33	Cloudy	6	9*	38	20
Gelcoat 39 PA	1.24	22	28	Cloudy mauvish	3	12†	39	21
Gelcoat 39	1.24	22	28	Cloudy	3 to 6	12*	39	21
Gelcoat 46 PA	1.40	18	22	Pinkish white opaque	3	9†	50	17
Gelcoat 47 PA	1.48	21	20	Pinkish opaque	3	9†	48	18
Gelcoat 65 PA	1.11	19	36	Cloudy mauvish	6	9†	42	17
Gelcoat 65	1.11	19	36	Cloudy	6	9*	42	17
Gelcoat 68 PA	1.10	18	42	Cloudy	3	12†	42	17
Gelcoat 68	1.10	18	42	Cloudy	3	12*	40	15
Gelcoat 69 PA	1.10	13	39	Cloudy	6	9†	40	15

Test methods and tolerances as in BS 3532 : 1962
All Crystic gelcoat resins are thixotropic.
* Resin, 100 pbw, Catalyst M, 2 pbw, Accelerator E, 4 pbw
† Resin, 100 pbw, Catalyst M, 2 pbw

*When pre-accelerated Crystic gelcoats are supplied coloured, the shelf life may be shorter than that of a normal pre-accelerated gelcoat. The resin should, if possible, be used within the recommended time indicated on the container

resin manufacturer, simple tests should be carried out, as previously mentioned, to ensure that no colour change due to interaction occurs.

In the early days of GRP body construction the obvious advantages of using a coloured gel coat was welcomed by laminators but after a number of failures to ensure a colour match between abutting panels, such as doors, boot and bonnet lids due to small variations in moulding conditions, most constructors dropped the idea and painted the mouldings in the conventional manner.

2.7 FILLERS

In the early days of GRP moulding fillers gained a poor reputation mainly due to the use of crudely ground limestone materials and to the misplaced enthusiasm with which they were applied. The use of an inert powder in a polyester resin gives both 'bulk' and through colour. Its use can also increase the compressive strength of the moulding provided the proportion used is not excessive.

The most important requirement of a filler is that it be completely inert, have no inhibiting effect on the resin and not be subject to deterioration or ageing under conditions which the moulding is to be used. Today a number of mineral fillers are available which fulfil these requirements and are specially prepared for use in conjunction with automobile body resins. It has now been recognised, however, that the advantage of using a filler stem from the improvement that can be gained in the physical properties of the laminate rather than from the cost savings that they produce. Preferred types include treated calcium carbonate fillers, particularly crystalline types. Other fillers used with polyesters include glass microspheres, which range in size from 20–500 μm, and wood flour. The latter it should be said finds limited use in automobile bodywork, being used largely in the furniture industry.

As a general rule filler content should be maintained as low as possible and finely divided filler should not be used at more than 25% of resin weight.

3

Materials: types of reinforcement

3.1 INTRODUCTION

For most applications in automobile bodybuilding polyester resins are used in conjunction with a reinforcement of glass fibres which, in addition to high strength, have excellent dimensional stability and are non-combustible. Their cost effectiveness also makes them suitable for most commercial applications. In recent years carbon fibres produced by controlled pyrolysis of organic fibres at very high temperatures have been used as reinforcement. They exhibit very high tensile strengths and modulus of rigidity but have been used in only a few applications because of their high cost. They have also been used in conjunction with glass fibres in mouldings which are required to combine high strength with lightness. Most glass fibres used in laminated structures such as automobile body parts are produced from a low-alkali glass, known as 'E' glass. Commercial glass fibres are drawn from molten glass as filaments of from 9 to 15 μm diameter. Before being wound or converted into the various forms in which it is used, the fibres are dressed with an emulsion designed to hold the fibres together to form a strand and to protect them during processing. The dressing also carries a coupling agent which promotes adhesion between the glass fibre and the laminating resin and imparts suitable properties to the surface of the glass.

3.2 PRODUCTION OF GLASS FIBRES

Most glass fibres used in the GRP industry are of the continuous type. In their manufacture glass spheres are heated in a platinum crucible or bushing as it is known in the industry. The molten glass flows out of holes, usually about 200, in the bottom of the bushing, each hole forming the source of one filament. By the use of a high-speed mechanised draw-off, filaments of a diameter down to 5 μm can be produced. After the filaments leave the bushing they immediately cool to a solid and on passing the winding head are brought together at a

Table 3.1   APPROXIMATE LAMINATE THICKNESS RELATIVE TO RESIN/
GLASS RATIO AND NUMBER OF PLIES (COURTESY BRITISH INDUSTRIAL
PLASTICS LTD.)

	Resin ratio by mass					
	3:1	2.5:1	2.25:1	2:1	1.5:1	1:1
	Resin content by mass (%)					
	75.0	71.4	69.2	66.7	60.0	50.0
Mass of mat per unit area (g/m^2)	Thickness (mm)					
300	0.87	0.75	0.69	0.62	0.50	0.37
450	1.31	1.13	1.03	0.94	0.74	0.56
600	1.75	1.50	1.37	1.25	0.99	0.74
750	2.18	1.87	1.72	1.56	1.24	0.93
900	2.63	2.25	2.06	1.87	1.49	1.11
1200	3.50	3.00	2.75	2.50	1.98	1.48

sizing station where a binder is applied to bind the filaments together
to form a strand. The count of the glass fibre is designated as the
number of 100 yard hanks/pound. The strands are supplied in counts
of 150, 225, 450 and 900.

3.2.1   Sizing

As mentioned, a size is applied to the continuous glass filaments
during production and is an operation of vital importance to the
performance of the end product. The sizes applied as liquids both
bind the fibres during processing and have specific tasks to perform
depending on the use of the reinforcement. It must not only protect
the filaments from abrasion whilst it is being drawn and processed but
must also contribute to the moulding process. Filaments destined for
spray-up roving will use a hard size whilst those to be used for
weaving will have a softer size applied. In all cases the size used will
contain a keying agent which is compatible with the polyester resin
and will promote adhesion between the two, i.e. a film former and
lubricant system. The keying or coupling agent is usually an organic
silicone compound: the film former, usually a polymer in emulsion
form such as polyvinyl acetate, and the lubricant usually based on
acid amides. In most cases other constituents are included to give
specific properties such as anti-static characteristics.

The performance of the size is of particular importance in the
production of glass fibres which are used in sheet moulding compounds (SMC) and dough moulding compounds (DMC). Leading
companies have many standard formulations of size which have been

Fig. 3.1. *Continuous strand roving.* (*Courtesy Scott Bader Co. Ltd*)

developed specially to relate to automotive requirements. Here an automobile manufacturer's own testing procedures have to be met with a GRP laminate and where in the main its required performance will be directly linked to the particular size used on the fibres.

Glass fibre reinforcement is available commercially as roving, chopped-strand random mat, needled mat, surfacing mat, continuous strand mat, combination mat, woven roving, woven cloth and tape, sheet moulding compound and dough moulding compound.

3.2.2 Roving

Continuous strand roving (Figure 3.1) is a product of several strands of glass fibre which are wound together without imparting a deliberate twist. Roving can be obtained with surface treatments for use with different resin systems and for different purposes. The material provides good overall processing characteristics including rapid penetration of the resin into the strands. They can be cut cleanly and disperse evenly throughout the matrix when moulding by the spray-up and premix methods. Rovings are supplied in the form of parallel or tapered 'cheeses' wound on a hollow core.

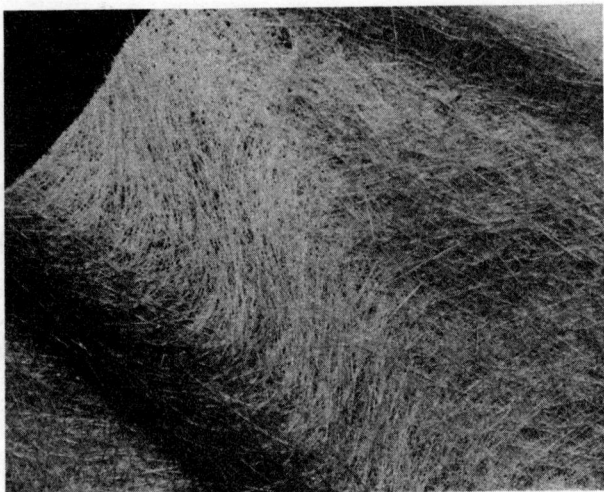

Fig. 3.2.   *Chopped strand mat.* (*Courtesy Scott Bader Co. Ltd*)

3.2.3   Chopped-strand mat (random mat)

This is the most widely used form of glass fibre reinforcement. It consists of bundles of strands chopped into lengths of approximately 50 mm and distributed randomly to form a mat which is held together by a binder to facilitate handling (Figure 3.2). The binder is usually polyvinyl acetate emulsion or a polyester powder, both of which are soluble in the laminating resin, and is selected to be of suitable strength to hold the fibres during processing. Also it must not catalyse or inhibit the cure or cause discolouration of the laminating resin. These mats are available in weights ranging from 225–1200 g/m² and in three different solubility types.

High solubility binder mat is used widely where the binder must remain soluble throughout the curing period of the resin as in translucent work and certain hand lay-up applications. Medium solubility types are used mainly for hand lay-up work where a certain strength retention in the mat facilitates impregnation. Low solubility mat is used largely in pressure moulding applications where excessive movement of the fibres must be restricted. The thickness of the laminate produced depends on the resin/glass ratio as well as the number of plies laid up. An approximation of laminate thickness relative to resin/glass ratio and number of plies used can be obtained by reference to Table 3.1.

Fig. 3.3.   Surfacing mat. (Courtesy Scott Bader Co. Ltd)

3.2.4   Needled mat

This type of mat is similar to chopped-strand mat but the fibres are held together by a mechanical needling process rather than by a chemical binder. Various types are available, each intended for a specific moulding method, for example, vacuum bag moulding and press moulding. Needled mats have the advantage of good drape property and wet-out rapidly giving good wet strength retention.

3.2.5   Surfacing mat or veil mat

These ultra lightweight mats (Figure 3.3) are used next to the gel coat to provide a good surface finish and to block out the fibre pattern of the underlying chopped strand, or other laminating reinforcement. Surfacing mat can also be used as a final lamination on the inner surface of a hand lay-up or spray-up moulding to give a smoother finish than can be obtained on the main reinforcement. These mats incorporate a resins soluble binder in the same way as chopped strand mats but are usually made of continuous fibres laid in a random pattern. Their thickness varies from 0.32–0.29 mm.

3.2.6 Continuous strand mat

This type of mat is used largely in resin injection systems and in hot and cold press moulding where the fibres can be displaced by the washing action of the resin. It is composed of continuous fibres disposed in a random fashion and is somewhat more expensive and bulky weight-for-weight than the more widely used chopped strand type.

3.2.7 Combination mat

These mats consist of one ply of woven roving which is chemically bonded to chopped-strand mat. Advantages of combination mats are that they form a strong and easily drapable reinforcement which offers the bidirectional properties of woven roving with the multi-directional characteristics of chopped-strand mats. The use of these mats can result in time saving in lay-up as two layers can be laid up in a single operation. In addition to the woven roving/chopped-strand combination others are available which combine multi-layer characteristics with surface finishing properties.

3.2.8 Woven roving

Woven rovings (Figure 3.4) provide the laminator with a heavy, easily drapable reinforcement which is available in a variety of widths, thicknesses and strengths. Rovings may be woven to form cloths in weights varying from 200–920 g/m². Most are woven with the same quantity of reinforcement in directions at 90° to each other although unidirectional woven rovings are also available for specific purposes. In general woven rovings are less costly than conventional woven fabrics.

3.2.9 Bi-directional roving

A new combination of woven roving and a chopped deposit is now available for hand lay-up. The bi-directional material reduces the crimp factor to a minimum, resulting in an ultimate tensile strength rating of 100%. The roving also allows control of the resin/glass ratio and prevents misalignment of the woven roving. Manufacturer's tests show that the reinforcement improves laminate strength by up to 40% . Figures 3.5 compares bi-directional with conventional material.

Fig. 3.4. *Woven roving. (Courtesy Scott Bader Co. Ltd)*

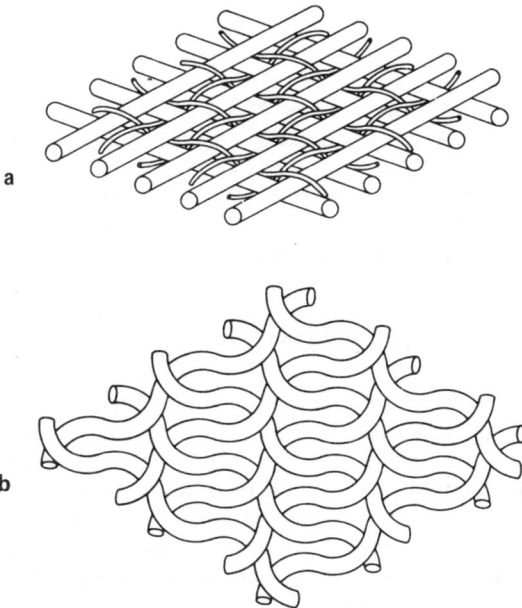

a

b

Fig. 3.5. *(a) Bi-directional woven roving compared with (b) conventional woven roving*

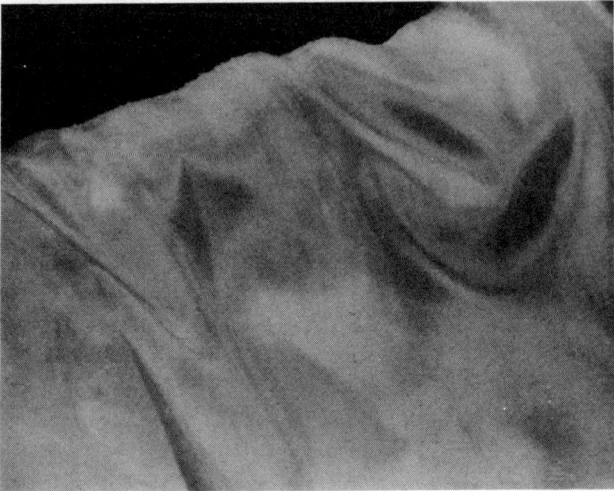

Fig. 3.6.   Woven cloth. (Courtesy Scott Bader Co. Ltd)

3.2.10   Woven cloths and tape

These cloths, Figure 3.6, vary widely in weight, thickness and weave. They are the most expensive form of glass reinforcement and do not find wide application in automobile work. Weights vary from 60–1130 g/m² and thickness from 0.90–1.2 mm. Various strength orientations are available and cloths can be obtained with various finishes, i.e. loomstate, heat-cleaned, Volanised and Garan treated, according to the application and resin formulation to be used. Woven tapes are also available in a variety of weights and widths for local strengthening applications and for duct winding.

3.3   SHEET MOULDING COMPOUND (SMC)

SMC is a combination of glass fibre reinforcement dispersed in a thermosetting polyester resin paste. A number of other materials are incorporated to provide desirable processing and moulding characteristics and optimum physical and mechanical properties.

Polyester resins used in SMC offer many advantages: ease of handling, rapid cure, good balance of mechanical and chemical properties, good dimensional stability, easily modified for special characteristics, and low cost.

Variations can be made in the composition of the base polyester to

Table 3.2 FORMULATION OF A GENERAL PURPOSE SMC RESIN PASTE
(COURTESY OWENS CORNING FIBERGLAS)

Ingredients			Parts	(%)
(1)	Polyester resin	OCF E-600	100	37.1
(2)	Catalyst	Tertiary-butyl perbenzoate	1	0.4
(3)	Mould release	Zinc stearate	4	1.5
(4)	Polyethylene powder	US Industrial Chemicals Co. Microthene* powdered polyethylene	6	2.0
(5)	Filler	Calcium carbonate	150	55.6
(6)	Pigment dispersion	–	8	3.0
(7)	Thickener	Magnesium oxide	1	0.4
		Total	270	100.0

Note: Mix sequence numbered at left. (4) and (6) are optional dependent upon application requirements

yield resin pastes with a wide range of properties during and after polymerization. Through selection and control of the ingredients, the polyester resin producer can synthesize a suitable material for specific end-use applications. The basic ingredients for polyesters used in SMC and the properties they impart in the formulations are presented in Table 3.2.

3.3.1 Production

After selection of the polyester resin base, a number of ingredients are added to produce a useful SMC resin paste. Additives used are catalysts, fillers, thickeners, release agents, pigments, thermoplastics polymers, polyethylene powders, flame retardants and ultra-violet absorbers. These components are mixed by the SMC manufacturer to exact proportions of the resin paste formulation. Some ingredients, such as release agents and thermoplastic syrups, can be added by the resin supplier. Each of the additives provide important properties to SMC during the processing and moulding steps and/or in the finished parts.

3.3.2 Catalysts

As discussed in Section 2.2 the primary purpose of a catalyst is to initiate the chemical reaction (copolymerization) of the unsaturated polyester and monomer ingredients from a liquid to a solid state.

Heat from the mould will cause the catalyst to decompose. This activates the monomer and polyester to form crosslinked thermosetting polymers. Generally, the addition of 0.3–1.5% by weight of catalytic agents will adequately promote the crosslinking reaction.

Organic peroxides are the principal catalysts used but as they vary in their ability to promote cure the particular catalyst selected for a specific resin formulation will depend upon one or a combination of factors, ie temperature of cure, rate of cure desired, means of activating catalyst (heat), cost, mass of resin involved, inhibitor type, and shelf life required.

The temperature at which the curing process is to be carried out usually determines the selection of a catalyst. For any given resin/catalyst system there is an optimum temperature at which peroxide decomposition initiates the monomer-resin polymerization process. Since SMC is usually moulded at temperatures of 132–165°C catalysts which are most effective as polymerization initiators over this temperature range are recommended.

Faster curing times for SMC polyester resins become an increasingly important factor in high volume moulding applications. Single peroxide catalyst systems, such as tertiary-butyl perbenzoate (TBPB), have been used successfully in low-profile SMC formulations to produce cure times as low as two minutes when moulding at tmeperatures of 150–160°C. Further reduction in curing times have been accomplished by combining the more stable peroxide, TBPB, with a faster, but less stable, curing peroxide such as t-butyl peroxyoctoate.

This dual catalyst system has been used in SMC resin formulations to produce curing times of about 90 s at moulding temperatures of 154.4–160°C. This is only one of several peroxide combination systems being investigated to provide high volume moulders with SMC materials having the lowest possible curing times.

3.3.3 Fillers

Fillers enhance the appearance of moulded parts, promote flow of the glass reinforcement during the moulding cycle and reduce the overall cost of the compound. Commonly-used fillers for SMC resin paste include calcium carbonate, hydrated alumina and clay.

Calcium carbonates are readily available fillers which can be added to polyester resin in large amounts, while still maintaining a processable paste (25–50,000 centipoise). They assist in reducing shrinkage of the moulded parts and in distributing glass reinforcement for better strength uniformity.

Hydrated alumina fillers provide flame retardance and are used largely in electrical appliance mouldings. Kaolin clays are sometimes used in combination with calcium carbonates or hydrated aluminas. When added at 10–20% of the total filler weight, the clays serve to control viscosity in the resin paste, to promote flow and to improve resistance to cracking in moulded parts.

The particle size of fillers is an important consideration for optimizing SMC formulations. The smaller the particle size, the better flow of the moulding material and the improved surface appearance of moulded parts. In some applications, there is evidence that coarser particles may have a degrading effect on glass reinforcements. Thus, in critical strength requirements, the finer particles would be preferable. Studies have shown that a combination of coarse and fine particle size fillers is frequently necessary for viscosity control when moulding.

3.3.4 Thickeners

Thickening agents include calcium and magnesium oxides and hydroxides. Thickeners initiate the reaction which transforms the mixture of SMC ingredients into a moulding material which is easy to handle and reproducible.

The SMC resin formulation usually contains 1–3% thickener. It is the final ingredient added to the resin mix, and it begins the chemical thickening process immediately.

The thickening reaction must be

(1) slow enough to allow wet-out and impregnation of the glass reinforcement

(2) fast enough to allow the handling required by moulding operations, as soon as possible after the impregnation step, in order to keep storage inventories low

(3) must give a viscosity at moulding temperatures low enough to permit sufficient flow to fill out the mould at reasonable moulding pressures

(4) must give a viscosity at moulding temperatures high enough to carry the glass reinofrcement along with the resin paste as it flows into the mould

(5) must level off in the mouldable range to give a long, useful shelf life and

(6) must be reproducible from run to run.

3.3.5 Release agents

Mould release agents are common components of SMC formulations. They are selected on the basis of their melting points being just below that of the moulding temperature. In theory, the release agent at the moulding compound–mould surface interface melts upon contact and forms a barrier against adhesion.

Commonly-used internal release agents for SMC include zinc sterate, calcium stearate and stearic acid. Stearic acid should only be used if moulding temperatures are below 127°C. Zinc stearate has a melting point of 133°C and can be used at moulding temperatures up to 155°C. Calcium stearate has a higher melting point, 150°C, and can be used at moulding temperatures up to 165°C.

Mould release agents should be used at the lowest concentrations possible to do an adequate job. Normally, they will be used in concentration less than 2% by weight of the total compound. Excessive amounts can reduce mechanical strength and cause blemishes on the moulded part surfaces.

3.3.6 Pigments

Pigments are supplied in two forms — as dry powders or paste dispersions. The paste dispersion pigment systems offer an advantage of less agglomerates in the resin paste. They can be added at lower concentrations than dry powders. Pigment concentration generally falls in the range of 1–5% by weight of the resin paste. Pigments can affect the cure time and shelf life stability of SMC systems. They may accelerate or inhibit the reactivity of the resin-catalyst system. Thus, pre-evaluation of the reactivity of a specific pigmented resin system is recommended.

The moulder must take into consideration the effect of colourants on end-product properties. Generally, pigments tend to slightly reduce the mechanical properties of a moulding. Samples of pigmented SMC should be tested against application specifications. Paint adhesion and bonding properties are affected very little, if at all, by pigments.

The original development of sheet moulding compound (now referred to as general purpose) posed no problem in the acceptance of most pigments to produce distinctive and decorative colours. Development of 'low shrink' and 'low profile' (see Section 3.4.3) SMC compounds improve many moulding characteristics and physical properties but reduce pigmentability. Low shrink compounds exhibit the minimum of contraction on curing. Low profile compounds are materials that cure with good surface profiles and show no waviness.

Generally, only black and off-white colourants are considered acceptable in low profile resin systems. However, low profile resin systems have been developed specifically for pigmentation.

3.3.7 Thermoplastic polymers

Thermoplastic polymers are combined with polyester resins to achieve low polymerization shrinkage in many SMC applications. Shrinkage is primarily controlled by varying the polyester/thermoplastic ratio. It is possible to attain near 'zero' shrinkage in moulded parts when thermoplastic polymers are added to polyester resins at concentrations of 40% by weight of the total resin system.

There are a number of thermoplastic additives that are compatible with polyester resins developed for SMC low shrink and low profile systems. Among those being used are: acrylics, polyvinyl acetate, styrene copolymers, PVC and PVC copolymers, cellulose acetate butyrate, polycaprolactones, thermoplastic polyester and polyethylene. The latter, in the form of powder of 8–30 μm, is a common additive of most general purpose and many low shrink SMC formulations. The addition improves surface smoothness of moulded parts, aids flow of the compound during moulding and does not cause colour problems. Powdered polyethylene is usually added to 2–5% by weight of the total resin formulation.

3.3.8 Flame retardants

Flame retardant additives such as hydrated alumina filler compounds normally satisfy most requirements; however, some of the more severe flammability classifications necessitate the use of flame retardant additives. These additives are used in conjunction with hydrated alumina fillers and halogenated polyester resins to provide maximum flammability performance. Flame retardant additives recommended for SMC are antimony trioxides, tris(2,3-dibromoproyl) phosphates, clorinated paraffins and zinc borates. Two of these additives are often combined at 1:1 ratios to offer more selective properties. They normally comprise about 3–5% of the SMC formulations.

3.3.9 Ultraviolet absorbers

Ultraviolet absorbers can be added to SMC resin blends when the moulded parts need to withstand extended exposure to sunlight. Generally, SMC resins are stabilized with approximately 0.1–0.25% of UV absorber of the benzotriazole or benzophenone type.

Table 3.3 A TYPICAL LOW-PROFILE SMC RESIN PASTE FORMULATION (COURTESY OWENS CORNING FIBERGLAS)

Ingredients			Parts	(%)
(1)	Polyester resin	OCF E-933	60.00	23.16
		OCF E-575	40.00	15.44
(2)	Catalyst(s)	1. Tertiary-butyl perbenzoate	0.50	0.18
		2. T-butyl peroxyoctoate	0.50	0.18
(3)	Mould release	Zinc stearate	4.00	1.50
(4)	Filler	Calcium carbonate	150.00	58.00
(5)	Thickener	Magnesium hydroxide	4.00	1.50
		Total	259.00	100.00

Note: Mix sequence numbered at left

3.4 FORMULATION OF DIFFERENT GRADES OF SMC

SMC can be supplied in many varieties of polyester resin paste/glass combinations to suit specific requirements. They can be categorized as one of three basic types or grades of materials, ie. general purpose, low shrink or low profile.

The most notable distinction between these three types of SMC is in the formulation of the resin systems. General purpose SMC has only unsaturated polyester resin as the base ingredient. Low shrink SMC may contain up to 30% thermoplastic polymer by weight of the total resin, while low profile SMC systems contains about 40% thermoplastic.

3.4.1 General purpose SMC

An example of a general purpose SMC resin formulation is shown in Table 3.2. It is representative of the polyester resin pastes recommended to produce SMC for a number of general purpose applications.

3.4.2 Low shrink SMC

Low shrink SMC applications generally require a two-component polyester resin system, utilizing the thermoplastic resin component at levels up to 30% to provide reduction in moulded part shrinkage. Table 3.3 presents a typical low shrink, two-component SMC resin formulation.

Table 3.4   A TYPICAL LOW-SHRINK, TWO-COMPONENT SMC RESIN
FORMULATION (COURTESY OWENS CORNING FIBERGLAS)

Ingredients			Parts	(%)
(1)	Polyester resin	OCF E-955	75	27.8
(2)	Thermoplastic component	OCF E-571	25	9.3
(3)	Catalyst	Tertiary-butyl perbenzoate	1	0.4
(4)	Mould release	Zinc stearate	4	1.5
(5)	Polyethylene powder	US Industrial Chemicals Co. Microthene* powdered polyethylene	6	2.0
(6)	Filler	Calcium carbonate	150	55.6
(7)	Pigment dispersion	–	8	3.0
(8)	Thickener	Magnesium oxide	1	0.4
		Total	270	100.0

Note: Mix sequence numbered at left. (5) and (7) are optional, dependent upon
application requirements

3.4.3   Low profile SMC

A typical low profile SMC resin paste formulation is shown in Table
3.4. It is identified as a two component system with the thermoplastic
component being added to the paste mix to enhance the surface
characteristics of the moulded part. Low profile resin systems were
originally developed for automotive applications where surface
smoothness and good paint adhesion were critical requirements of
the SMC moulded product.

3.4.4   Reinforcement

The development of sheet moulding compounds have made a notable
impact in automobile bodywork and, in many cases, are used for the
quantity production of components such as bumpers, headlamp
casings and front end panels. The Chevrolet Corvette body now uses
outer panels of SMC. Mouldings use matched metal tooling thus
greatly reducing the labour intensity problems associated with hand
or spray-up processes whilst offering higher strength/weight characteristics combined with consistency of dimensional tolerance.
   The strength characteristics of a SMC moulding are dependent not
only on the type of resin and filler used but also, and to a greater
extent, on the amount and type of glass reinforcement. Basically there
are four types of glass fibre structure available in the form of SMC. In

Table 3.5 TYPICAL PHYSICAL PROPERTIES OF GRP WITH DIFFERENT TYPES OF GLASS FIBRE REINFORCEMENT (COURTESY SCOTT BADER LTD.)

Properties	Unit	Chopped strand mat	Woven rovings	Satin weave cloth	Continuous rovings
Glass content	% weight	30	45	55	70
	% volume	18	29	38	54
Specific gravity		1.4	1.6	1.7	1.9
Tensile strength	MPa	100	250	300	800
Tensile modulus	GPa	8	15	15	40
Compressive strength	MPa	150	150	250	350
Bend strength	MPa	150	250	400	1000
Modulus in bend	GPa	7	15	15	40
Impact strength, Izod, unnotched*	kJ/m²	75	125	150	250
Coefficient of linear expansion	x10−6/°C	30	15	12	10
Thermal conductivity	W/m K	0.20	0.24	0.28	0.29

*Tested edgewise

Weight of resin in pounds   Weight of additive in grammes

%of additive required

To use the nomograph place a straight edge from the 'Weight of Resin' scale to the '% of Additive' scale. The intersection of the straight edge and the 'Weight of Additive' scale shows the weight in grammes of the additive required. When using liquid additives such as Catalyst or Accelerator it may be assumed that the weight of additive in grammes is equivalent to the volume of additive in ml. or cc's.

Example A

Weight of resin is 5 lb and 4% additive is required. The amount of additive necessary is 91 g.

Example B

Weight of resin is 10 lb and 1% additive is required. The amount of additive necessary is 45 g.

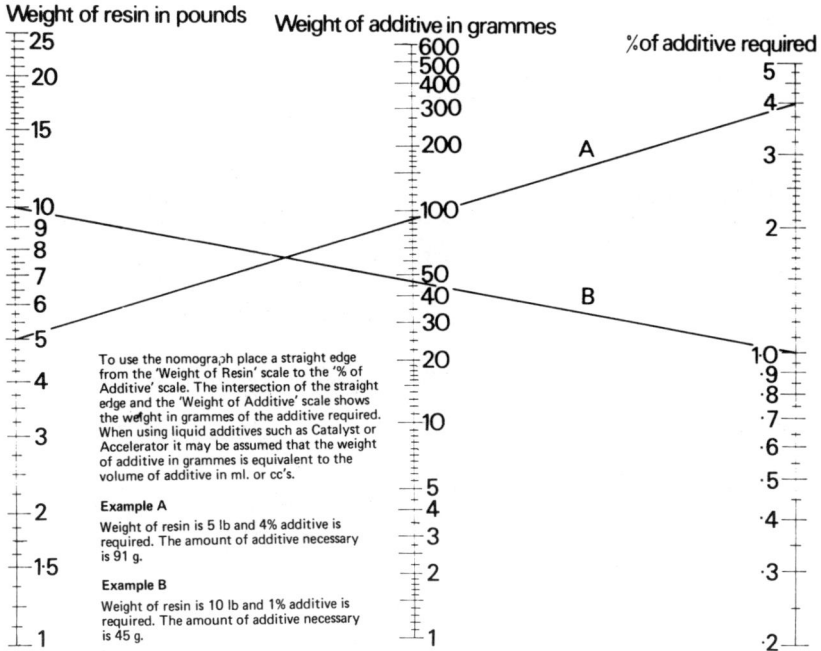

Fig. 3.7.   Nomographs giving weight of resin/additive

the first, the chopped rovings are 25 mm and 52 mm long and are distributed approximately in a random pattern to give equal strength in all directions. This type permits the production of complex mouldings with a good surface finish and with a glass content of between 25% and 35%. A second type incorporates endless rovings which are arranged parallel and longitudinal to chopped fibres, so greatly increasing the strength of a component in the direction of orientation. In a third type, rovings 100–300 mm in length are laid parallel but staggered, a form which provides better flow during moulding than the second configuration. Lastly, a wound formation of crossed roving tapes laid at 20° can be used to obtain high directional strengths. This type is more costly than conventional types but permits a glass content of between 65% and 75% to be achieved for specialised applications.

3.5   DOUGH MOULDING COMPOUNDS

Dough moulding compounds are produced by mixing a suitably catalysed polyester resin, mineral filler, pigment and glass fibre

Glass weight g/m^2

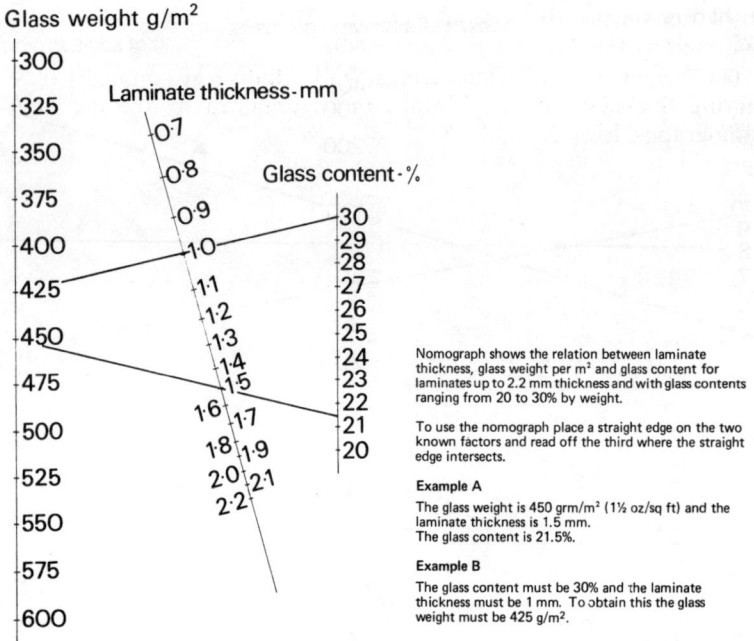

Nomograph shows the relation between laminate thickness, glass weight per m^2 and glass content for laminates up to 2.2 mm thickness and with glass contents ranging from 20 to 30% by weight.

To use the nomograph place a straight edge on the two known factors and read off the third where the straight edge intersects.

Example A

The glass weight is 450 grm/m^2 (1½ oz/sq ft) and the laminate thickness is 1.5 mm.
The glass content is 21.5%.

Example B

The glass content must be 30% and the laminate thickness must be 1 mm. To obtain this the glass weight must be 425 g/m^2.

Fig. 3.8. *Nomographs giving glass weight/laminate thickness*

reinforcement to form a dough-like material which can be used to mould components in matched metal moulds. As would be expected some breakdown of the glass fibres occurs in preparation of the material which results in some reduction in strength as compared with materials containing long fibres, but the loss of strength is acceptable for many ancillary components of a body which are not subject to high stress.

3.6 STRENGTH OF GRP LAMINATES WITH DIFFERENT REINFORCEMENT

The physical properties important to the automobile body designer are dependent on a number of factors. These include the type of reinforcement used, the glass/resin ratio obtainable, which will be governed in hand and spray lay-up processes largely by the skill of the operator, and in other moulding processes by the designer and the equipment used.

As described briefly in the foregoing there is a wide range of glass fibre reinforcement materials available and as a guide to the physical properties that can be obtained Table 3.5 lists typical properties conferred by the different types discussed.

The weight of resin reinforcement and additive in laminates of varying thickness up to 2.2 mm can be obtained by the use of nomographs, Figs. 3.7 and 3.8.

4

Moulding methods

4.1 HAND LAY-UP OR CONTACT MOULDING

As described briefly in Chapter 1, this method is both the oldest and the simplest for moulding glass reinforced polyester resin structures. It can also be used for making the mould itself from a master pattern. Whether the mould is male or female depends on the component itself, but the overriding factor is that the surface in contact with the mould will have a smooth finish whilst the other will be rough. The actual degree of roughness will depend very largely on the type of reinforcement used and the skill of the operator, but this is not of great importance in body mouldings as the interior is usually fully trimmed. A further factor, and one of particular importance in bodywork which is exposed continuously to weathering, is that the surface in contact with the mould can be given a thin coating of unfilled resin. This layer, known as the 'gel' coat, not only masks the reinforcing fibres but improves resistance to weathering and corrosive attack.

4.1.1 Shop conditions: safety measures

In many ways the inherent simplicity of the hand lay-up process for producing moulds inexpensively and for moulding large complex shaped components, coupled with the wide range of working conditions tolerated by polyester resins, has contributed to abuse by inexperienced and misinformed moulders. While quite passable results on small simple shapes can be obtained under adverse conditions, consistency, optimum strength and surface finish can only be ensured by carefully controlled conditions at each stage of the process. The moulding shop should be free from dust and draughts and the temperature should at all times be maintained at a minimum of 20°C, with low humidity. All resin materials must be stored in a cool dry area and the various components kept well apart until required for use. Glass reinforcement in particular should be kept dry and preferably warm as any moisture present in the fibres will inhibit impregnation by the resin and result in soft uncured areas in the finished laminate (Sections 9.1.15, 9.1.16). Various equipment, detailed in Chapter 11, is now available for handling, measuring and mixing the components of the resin, for rolling and consolidating the laminate and for trimming the finished moulding. Regulations now cover the use of styrene emitting resins, while common sense will dictate the use of dust masks when sawing or sanding cured mouldings.

4.1.2 Safety precautions[2]

Polyester resins are flammable and most have a flash point around 30°C. The storage life of unpigmented polyesters varies from 6–12 months at below 20°C when stored in the dark in metal drums. If the temperature exceeds 20°C, even for a few days, or the resin is exposed to daylight its storage life will be greatly shortened. Accelerators are also flammable and should also be kept below 20°C at which temperature they will have a shelf life of about 6 months. Catalysts are a special hazard and must be kept cool. Catalysts and accelerators must never be brought into contact with each other as they react with explosive force.

There are certain precautions that should also be observed in handling resins, catalysts and accelerators. As most polyesters contain monomeric styrene which is a strong grease solvent, resins should not be allowed to remain on the hands as it can cause irritation. To most operators, rubber gloves are unsatisfactory and the best alternative is the use of a barrier cream. While resin can be removed from the hands with acetone, this will also dry the skin and may cause dermatitis. Catalysts are also very harmful to the skin and direct contact should be avoided. If handled it is vital to wash immediately in warm water.

4.2 MASTER PATTERN

A detailed description of the construction of master patterns and moulds are given in Chapter 10. However, before discussing the hand-lay-up process a brief outline of the requirements of the master pattern and mould will be given in the order that the hand-lay-up process can be followed without difficulty.

Essentially, the master pattern is a replica of the component to be moulded. In cases where a metal panel or moulding exists, it can be made from this, although it is more usual to work from a wood or plaster model. Factors of impórtance in the design of the pattern are that the mould must be readily removable when completed and that the surface finish on the master must be as perfect as possible, as any imperfections will be reproduced on each component moulded. Many of the problems met with in the early days of hand lay-up moulding stemmed from insufficient thought being given to release angles in the mould, which caused delays in production and often led to mould damage. Where possible a release angle of not less than $1\frac{1}{2}°$ is required on vertical surfaces. Undercuts should be avoided as they complicate release, while sharp angles and corners lead to air pockets in the finished moulding. In automobile bodywork most configurations are such that the mould must be made in sections in order to be able to remove the finished moulding in one piece.

Whatever material is used for constructing the master pattern it must be absolutely rigid and its surface must be adequately filled by the application of a hard compound such a furane resin which, when cured, can be sanded and polished to give a high surface finish. Attention to the excellence of the master cannot be too highly stressed and any expenditure of time at this stage will be amply repaid when the mould is in production.

4.2.1 Laying-up the master pattern (mould construction)

Before any resin is applied to the master its entire surface is waxed and coated, ideally by spray, with a release agent such as polyvinyl alcohol in order to ensure clean removal of the finished mould. When dry, a thin spray coating of catalysed and accelerated resin is applied which forms a gel coat. This gives a mould surface capable of reproducing the contours of the pattern accurately and free of reinforcement. After curing, the gel coat is covered with a layer of special thin surfacing mat and general purpose resin. When this layer has been allowed to cure a layer of chopped strand mat and resin is added and consolidated by rolling with special rollers. After about 24 hours the required thickness of the mould can be built up with successive layers of chopped strand mat and resin, each layer being consolidated by rolling. Timing will depend on the resin formulation used.

To ensure that the mould will retain the exact shape of the pattern and will not distort in use, stiffening is incorporated where necessary. This can consist of shaped wooden battens, plastic hose, rigid foam section or similar formers which are overmoulded to the exterior surface of the mould, again using chopped strand mat or for greater strength, woven rovings (Figure 4.1). This operation completes the

Fig. 4.1. Alternative methods of mould stiffening. (Figures 4.1–4.9 courtesy Scott Bader Co. Ltd)

laying-up of the mould which should remain in place on the master for about two days. When fully cured it can be removed, an operation often facilitated by the introduction of water between the two surfaces to dissolve the polyvinyl alcohol parting agent. The moulding surface is then inspected for pin holes or imperfections after which it is wax polished before putting into service.

4.2.2 The moulding operation: applying the gel coat

The wax used for polishing out the mould before use should be silicone-free. Silicones are excellent parting agents but it is very difficult to remove all traces of the wax and its presence will inhibit subsequent painting operations. Usually it is necessary to carry out three or four polishing operations on a new mould and for the first few mouldings it is advisable to use a release agent such as polyvinyl alcohol to ensure good release (Figure 4.2). With most moulds this treatment is only necessary when producing the first few mouldings and once the mould has become 'run-in' the application of alcohol type release agent can be discontinued providing that the mould is rewaxed and the surface polished at regular intervals. Most resin manufacturers supply special 'self-release' gel coat resins which can be used in applications where the moulding does not require to have a highly polished surface.

When the PVA release agent is thoroughly dry, a process which can

Fig. 4.2.   Application of release agent to mould surface

be accelerated by the use of a low-temperature hot-air blower, the gel coat can be applied (Figure 4.3). This coat which forms the outer surface of the moulding consists of a comparatively thin layer of unreinforced and somewhat flexible resin. It serves a number of purposes; first, providing a smooth unbroken skin preventing the protrusion of glass fibres, second, it forms a resilient coating which can be pigmented and third, it provides protection against weathering and corrosion.

Colours are available as dispersion in a suitable polyester resin and can be added in a proportion of up to 10% by weight to the gel coat resin. Mixing must be thorough to ensure complete distribution, care being taken that no air is introduced which would result in the formation of small pin holes in the surface of the moulding. For this reason hand or slow speed mixing equipment should be used.

Most propriety gel coat resins are pre-accelerated and can have thixotropic characteristics which allows the resin to be brushed or sprayed. If left undisturbed on the mould surface, it thickens and can then be applied to inclined or vertical surfaces without danger of an increase in film thickness in the lower areas of the mould due to drainage. For brush applied gel coats the catalyst is dispersed in the gel coat resin immediately before use. The quantity should be accurately weighed out to give a bulk gel time of about 15 minutes and mixed in the resin in the same way as the pigment. Once catalyst has been added to pre-accelerated resin it will have a specified gelling time and mixing should be carried out as quickly as possible. For this

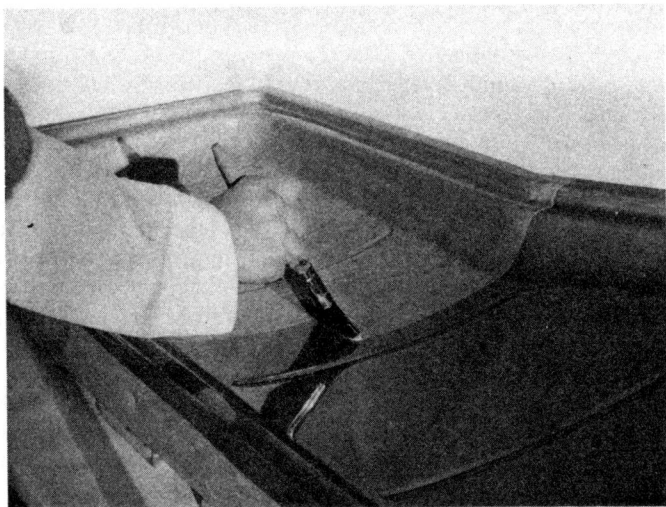

Fig. 4.3. Application of gel coat

reason a mechanical mixer which does not produce a vortex should be
used.

The thickness of the gel coat should be approximately 0.4 mm for
automobile work giving, for calculation purposes, a coating weight of
about 600 g/m². Although the gel coat can be applied by brush or by
spraying, the latter method is recommended wherever possible as it
gives a more even coat. Also if the mould is of a large area it permits
the operation to be completed without any problems arising from the
comparatively short pot life of the resin. When spraying the gel coat,
catalyst is not added to the resin but is injected into the resin stream as
it emerges from the spray gun. The technique and equipment are
described in Chapters 4 and 11. Under suitable shop conditions of low
humidity and ambient temperature of not less than 16°C, the gel coat
will be ready for lay-up approximately $1\frac{1}{2}$ to 2 hours after application
depending on the configuration of the mould.

4.2.3 Laying-up the mould

As detailed in Chapter 3 there are a number of different types of
reinforcement that can be used for hand lay-up work but whether the
choice is chopped strand needled or continuous strand mat, woven
rovings or cloth the sequence of operations is basically similar. In
some cases the main reinforcement is layed-up directly on the

partially cured gel coat but the more general and safest practice, in order to avoid any danger of air voids between the laminate and the gel coat, is to use a first layer of glass surfacing tissue which is easily impregnated and gives a clear indication of contact with the gel coat itself. The tissue is cut to the approximate size required and is pressed down, using a brush, on to a thin coating of general purpose laminating resin which is applied either by brush or spray. This resin and that used subsequently for the main body of the laminate is mixed with a proportion of catalyst and accelerator to give an adequate pot life before gelling occurs. The actual period will depend on a number of factors — the size of the mould, the weight of the reinforcement used and the skill and speed of the operator. Most laminators work on a pot life of at least $\frac{1}{2}$ hour but until experience is gained on a particular mould it is safer to work with somewhat longer gel times. It should be borne in mind that a bulk of resin will gel far more rapidly once the reaction has commenced than a thin layer in the mould due to the exothermic heat produced by the crosslinking process.

Once the surfacing tissue layer has partially cured, the laminating process proper can be commenced. For the first layer a chopped strand mat of comparatively light weight is used. As detailed in Section 3.2.3, the material is produced in several different weights, those commonly used being 1, $1\frac{1}{2}$, 2 and 3 oz/ft^2. The choice will depend very largely on the particular moulding and the stage of the laminating operation. However, the thinner mats are easier to impregnate and are less likely to produce air voids by bridging of identations in the mould or at mat edges by the glass fibres.

Laminating should always be commenced by applying a liberal coating of resin to the mould on to which the previously tailored mat is laid. Impregnation of the mat with resin is then carried out by the use of metal rollers which bring the resin up through the glass without leaving air bubbles (Figure 4.4). The preferred rollers (Section 11.2) consist of a bank of metal washers and these are used with a light pressure and short strokes at various angles across the glass mat. Paddle type rollers can be used at a later stage. Some laminators use lambswool rollers for applying the resin but these are not recommended for bringing up the resin in the early stage of impregnation.

Experience will dictate the amount of resin to apply before laying on a given thickness of mat but the aim should be to apply just sufficient to impregnate each layer (Figure 4.5). Ideally the resin/glass ratio should be 2:1 and 2.5:1.

Resin rich laminates, while being unnecessarily heavy and costly, are far weaker weight-for-weight than those with a high proportion of reinforcement. Where a coloured gel coat has been used it is advantageous to pigment the resin used in the first layer of glass mat

Fig. 4.4. Rolling of impregnated mat

to the same shade, although this is not usual practice in automobile work. In this way any thin areas of gel coat will not shown in the finished moulding.

Sharp corners and small radii, whether male or female, are to be avoided wherever possible in component design. In areas of a mould containing small sunken radii difficulty arises in getting chopped strand mat to remain in contact with the surface of the gel coat. This is due to the inability of the glass to retain a small radius bend but this, however, can be overcome by laying bundles of well impregnated roving along the areas to bring them up to the approximate level of the surrounding mould surface before the first layer of mat is applied (Figure 4.6). Great care should be exercised when laminating a moulding of this type because any voids caused by sunken radii will be virtually impossible to correct when the component has been removed from the mould.

When this layer of glass mat has been thoroughly impregnated and no dry white patches are visible, subsequent layers of resin and reinforcement are applied using the same methods but, if required, with heavier glass mats. It will be found when producing thick mouldings that there is a practical limit to the number of layers of glass mat that can be applied whilst the previous layer is still liquid. The problem is that a thick layer of reinforcement, fully impregnated with resin, will tend to move under the pressure of the washer roller and thus become unmanageable. Additionally, due to the exotherm

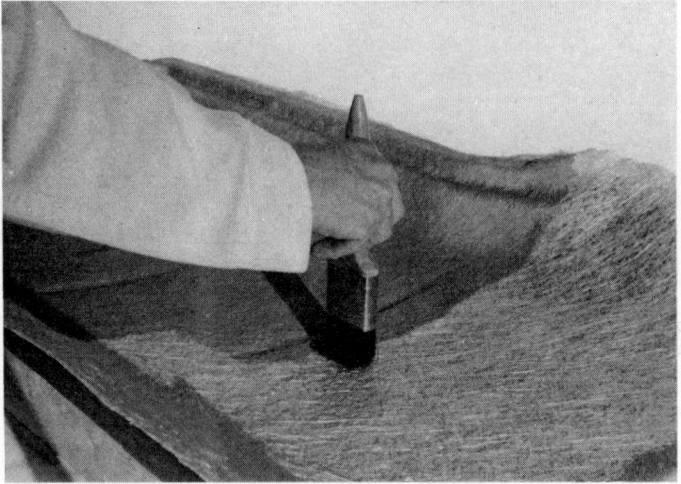

Fig. 4.5. Impregnating mat

Fig. 4.6. Method of filling sunken areas with roving prior to applying mat

produced during cure, considerable heat will be generated in a thick laminate which can warp or discolour the moulding and in extreme cases, damage the mould. At best the heat will destroy the wax coating. Thus, it is advisable when producing mouldings with more than say two layers of glass mat to allow the first layers to gel before proceeding. Cold cure polyesters require a considerable period for complete cure unless heated, but in order to avoid any danger of delamination subsequent layers must be applied without delay, before the cross-linking reaction is completed, otherwise subsequent layers will tend to delaminate. It is difficult to lay down an exact time between the application of successive layers as the ambient temperature will vary the gel time considerably, but providing the underlayers have gelled and are firm to the touch lamination can proceed with assurance that no delamination will occur. As a final process in the laying-up operation some laminators apply a surfacing tissue to the inner surface of the moulding. If carefully laid and impregnated, by

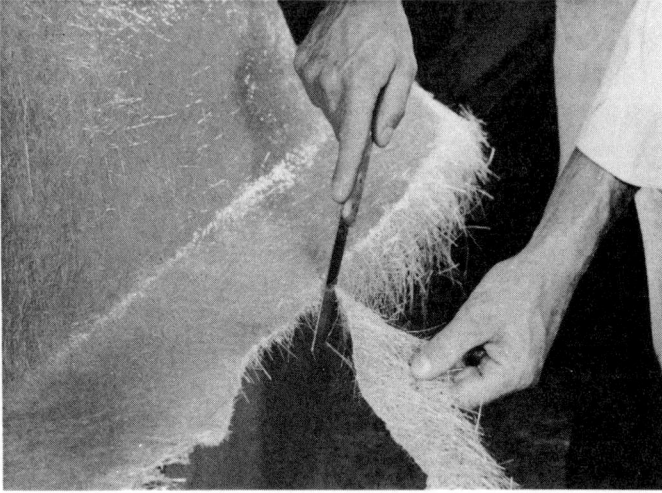

Fig. 4.7. Trimming the edge of a moulding at the 'green' cure stage

dabbing with a brush, the tissue mat will give a much improved finish in cases where the inner surface can be seen.

When laying up is complete but before the moulding has fully cured, the excess laminate can be trimmed back to the mould edge with a sharp knife (Figure 4.7). Where this operation presents difficulty the parameters of the mould can be scored into the moulding and the excess removed by fine toothed hacksaw or better an electric jig saw or disc cutter (Figures 4.8 and 4.9). These methods of trimming, while satisfactory for short run production items, are time consuming and where mouldings are produced in any quantity a more accurate and faster method is to use a trimming fixture. The fixture can be moulded off the original master pattern as a skeleton to house the moulding and should be provided with reinforced edges around the periphery to act as a guide for the saw and to give a datum line to which the edge of the fully cured moulding can be sanded with a high speed abrasive disc.

4.2.4   Demoulding

Whatever method is used for trimming, the moulding should remain undisturbed in the mould until all exothermic heat has been dissipated. The slight shrinkage which takes place on full cure will

Fig. 4.8.   *Alternative method of trimming a fully cured moulding with a power jig-saw*

facilitate its removal. Care must be taken in removing the component from the mould as either moulding or mould can easily be damaged by careless handling. One method which is used to facilitate the removal, particularly of large parts, is to mould a number of lifting points at rigid areas so that a number of operators can grasp the moulding simultaneously or it can be suspended by a hoist so that the mould drops away. Another method is to mould flanges on the edges of the component for the insertion of soft wedges.

Components which are thin in relation to their size are often demoulded by the use of vents built into a number of points in the mould. These are used to inject either air or water under pressure to assist in lifting the moulding. In difficult cases, several of these methods can be used simultaneously to demould components.

Precautions should also be taken to ensure that any cured laminate overhanging the mould does not interfere with ejection.

4.2.5   Cure and post cure: maturing

Final cure will vary with the type of resin and catalyst used, depending whether the component has been produced by hand lay-up or spray lay-up methods. Parts that are not required for use immediately are air cured while parts that are required for use soon after moulding are oven cured. In the first method, if the part can be stored at

Fig. 4.9. Trimming fully cured moulding with a diamond wheel.

temperature preferably of between 70°–80°F and the correct type of catalyst has been used, the part will attain a usable state of cure in storage. It should be noted that in any event the part must be sufficiently rigid to retain its form and not to distort after removal from the mould. Distortion will occur if demoulding is carried out too soon or in the case where oven curing is used, whilst the part is too hot.

As a guide, three changes in the laminate can be seen:
(1) There should be very little tack on the surface.
(2) Colour of the laminate will change from the wet state.
(3) There will be no exothermic heat.
Large components should be placed in a jig after removal from the mould and when in storage to ensure that no deformation occurs.

Several types of heating can be employed for 'oven curing' including convector type hot-air ovens or banks of infra-red lamps but if the latter are employed they should not be placed so close to the moulding that pattern heating occurs. As a check a Barcol hardness test of the surface of the laminate can be made when the part has been cooled. A reading of 40 or more will indicate that the part is near to final cure. Figure 2.1 gives equivalent post-curing times and temperatures for an average polyester resin system.

4.3   SPRAY-UP

The simultaneous deposition of resin and chopped glass fibre reinforcement into an open mould has permitted shorter working cycles in the laying up of large automobile panels and has resulted in a reduction in the pre-cutting and handling of reinforcements. The system also permits economies as it uses rovings which are the least costly of all forms of glass fibre reinforcement.

There are a number of systems commercially available, but basically each has the same function, to chop the roving into specified lengths, usually about 20–50 mm, and spray it into the mould at the same time adding catalysed resin in a controlled ratio.

In the early days when laminators realised that there were advantages to be gained by spraying resin and glass for open or contact mould work the only equipment available was the conventional, external atomising spray gun which was a modified version of the paint spray unit. These guns operate with dual nozzles and spray catalysed and accelerated resin in two streams each fed from pressure pots. In another version liquid catalyst is accurately metered into the pre-accelerated resin and mixed either in the gun or outside through a specially designed spray nozzle. In both systems a roving chopper delivers the glass fibre roving into the resin stream. In systems in which mixing takes place within the equipment through flushing is vital to eliminate gelling of the resin mix within the gun.

A more recent development is the 'airless' or hydraulic system in which a pump is used to generate the necessary fluid pressure for spraying. In this type of gun the resin streams are finely atomised at the nozzle and impinge on the mould at relatively low velocity thus reducing bounce back and overspray produced by the more conventional type of air atomising equipment. Additionally, these later guns are capable of handling all varieties of resin including filled and highly viscosity fire retardant types used in automobile work.

A further advantage in production work is speed of application and their ability to lay down high resin rates. They normally operate on an external mix of resin and catalyst with the latter readily imploding into the airless stream of resin to ensure a good mix without the entrapment of air which can occur when using the earlier type.

Application rates with this type of equipment range from 1200–1500 cm^3/min and with a suitable roving chopper will lay down up to 4.5 kg of glass/min.

4.3.1   Setting glass/resin ratio

The following procedure should be followed when spray laying-up a large area mould. The spray gun is adjusted to produce glass fibre in the range 25–30 mm long using continuous roving with or without a tracer. The use of a coloured tracer yarn can help the operator to maintain uniformity of reinforcement by providing a visual control. The ratio of glass to resin must be established at the required figure. Several methods of calibration can be used: the following is but one example designed to give approximate deposit of 1.36 kg/min of resin and glass fibre.

Operation

(1) Assume that an output of 0.45 kg/min through the gun using one-strand roving is required (one strand is used only as an example — multiple strands can be passed through the gun).

(2) Run one strand of roving through the gun (chopper) for 15 s collecting the output for weighing. Vary the chopping speed to give an output of glass at a rate of 115 g in 15 s.

(3) Assume that 30–33% by weight of glass is required in the finished laminate. The amount of resin required will be 0.90 kg/min based on 0.45 kg/min of glass output.

(4) Dispense prepared resin (catalysed, etc.) into a container for weighing. Dispense for 15 s and weigh the resin collected. Adjust the resin flow (the method will depend on the type of gun used). Correctly set, the output will be 225 g of resin in 15 s (0.9 kg/min), a total of 1.36 kg/min.

4.3.2 Spraying operation

It is recommended that in quantity production situations all flash should be removed during the spraying operation or while the resin is still soft in the same manner as shown in Figure 4.7. Difficult to fill areas should be dealt with first by spraying with a light coat of resin only, followed by glass and resin until the total thickness has been deposited. These areas should then be rolled out before proceeding to the remainder of the mould. Surfaces may be covered by one or two passes of the gun working along the length of the part and ensuring coverage of deposit from one pass to the next. Many operators prefer to make two passes rather than one claiming better thickness control and elimination of air inclusion, particularly when working on thick laminates.

When lay-up is complete, rolling with a washer type roller consolidates the laminate and ensures good contact with the gel coat in a similar manner to a hand lay-up operation. Rolling should commence at the centre of the mould working outwards. In areas where the reinforcement spans indentations in the mould, fibres are worked down from edge to centre. The thickness of the laminate can be checked using a fine pin and disc. The pin is lightly waxed and pressed through the laminate until the point is in contact with the mould surface. The process is carried out at a number of points. When the disc touches the laminate surface the thickness is correct. If necessary re-roll to fill the hole made by the pin. Curing and removal of the component from the mould follows the procedure detailed under 'Hand lay-up' in Sections 4.2.3 and 4.2.4.

4.4 VACUUM BAG MOULDING

The vacuum bag method of moulding is not so widely used as hand and spray-up processes. It offers advantages for certain types of component due to the higher glass/resin ratios that can be obtained and to the fact that an acceptably smooth finish can be obtained on the side not in contact with the mould surface. As the process is usually used when a higher rate of production than can be obtained by hand lay-up is required, some heating to accelerate cure is normally adopted. In this case the mould will be of metal and can be heated either by built-in steam or hot water ducts or, more conveniently, by resistance elements embedded in the body of the mould. One advantage of using the vacuum method is that the mould is stressed evenly over its entire surface and can be of comparatively light construction.

Depending on the configuration of the component, the bag can be (1) a diaphragm clamped around the periphery of the mould or (2) a complete bag which envelopes both the moulding and the mould. This arrangement is most suitable for small components and those which require to be male moulded, ie. with the 'mould-finish' on the inner surface. It will be appreciated that for method (2) hot curing is normally carried out in a medium temperature oven in order to overcome the problem of providing an internal heating system for a mould which is completely enveloped in a bag.

The bag can be of any material capable of being stretched and of withstanding the oven temperature when this method of curing is used. Rubber sheet provides the necessary extensibility but is attacked by resin and has a limited life at hot cure temperatures. Rubber can be kept out of contact with the laminate by the use of a cellophane or similar film but if available a sheet or bag of polyvinyl alcohol will offer superior properties both from the point of view of resin and heat resistance.

When using the vacuum bag method of moulding a sheet of porous material is interposed between the bag and the cellophane film to ensure that the air is evacuated evenly from the entire surface of the laminate. Resin traps are provided at each suction point to prevent resin passing into the vacuum line.

The method of impregnation and laying of of the laminate is the same as that of hand or spray lay-up but limited in application to the size of oven available.

4.5 PRESSURE BAG MOULDING

As described briefly in Section 1.6.4, the pressure bag method of moulding is similar to the vacuum method in that it employs a flexible membrane to exert pressure over the surface of the moulding. However, it has the advantage for certain types of high strength component, that the pressure available is not limited to atmospheric but, within limits, can be increased to allow high glass/resin ratios. Remarks made with regard to the type of bag used in vacuum bag moulding and method of clamping, etc. apply equally when the bag is pressurised. Laying up and impregnation methods are also similar as is the choice of ambient temperature, heated mould or oven curing. Similarly, a porous material is used to permit an even overall pressure on the laminate with resin reservoirs to accept excess resin as it is squeezed out of the reinforcement.

4.6 AUTOCLAVE MOULDING

So far as the author's experience is concerned this method of moulding finds little application in automobile bodywork but for completeness is mentioned as an alternative to the pressure bag process. One obvious advantage of the system is that with the availability of an autoclave a large number of small mouldings can be cured simultaneously with savings in both time and labour. However, an autoclave is an expensive and space consuming piece of equipment, comparatively costly to operate and therefore its application is usually confined to highly specialised types of moulding.

4.7 COLD PRESS MOULDING

Cold press moulding finds its major applications in the area between hand and spray lay-up processes and the faster but more expensive hot press moulding processes which have been developed for long production runs of smallish components. Pressures required for cold press moulding can be as low as 0.98 kgf/cm^2 and as the moulds are not required to be heated they can be made of GRP — usually incorporating epoxy resins rather than polyester — although the latter can be used. If comparatively long production runs are envisaged, some moulders use metal moulds or concrete moulds which preferably should be faced with a reinforced epoxy or a furane resin to give a smooth and non-porous surface.

Cycle times average 15–20 min and as with all matched mould processes components produced have 'mould finish' on both inner and outer surfaces. For long production runs it is simpler and more economic to mount the moulds in a low pressure hydraulic press (see Section 11.16.1). To ensure alignment of the mould halves, provision is made in the form of accurate platen guides or the incorporation in the mould itself of pillars and bushings. Any misalignment will result in variations in the wall thickness around the component and this will be more pronounced the deeper the draft of the part. It is usual practice to incorporate mould stops around the mould periphery to ensure that mouldings are of consistent wall thickness and that the desired glass/resin ratio is maintained. This is normally in the range of 2:1 to 5:1 depending on the properties required in the finished moulding. Sometimes where components of high surface finish are produced it is necessary to apply a gel coat to either, or both, male and female moulds. Gel coat formulations used for hand and spray-up mouldings are suitable but they must be allowed to gel before the mouldings operation is completed and thus will necessarily lengthen the production cycle.

For shorter runs or where components are comparatively small adequate results can be obtained by clamping the mould halves together using 'Speetog' type toggle-clamps or even bolts and wing-nuts. As on a press care must be taken to ensure mould alignment. This can be accomplished by the use of pins and bushes moulded into the flanged periphery of the mould and clear of any resin overflow which will occur on final closing. The number and size of the pins and bushes will naturally depend on the size of the component but they must be of sufficient length to ensure engagement well in advance of mould closure. Details of mould design including resins, the location of pinch-off and mould travel stops, etc. are shown in Figures 1.6–1.7.

The cold press method using GRP moulds has been applied successfully for runs of up to 2000 using two-component general purpose resin or two component fire-resistant resin systems. Both are fast gelling and give acceptable production outputs.

In operation the glass reinforcement and resin are introduced

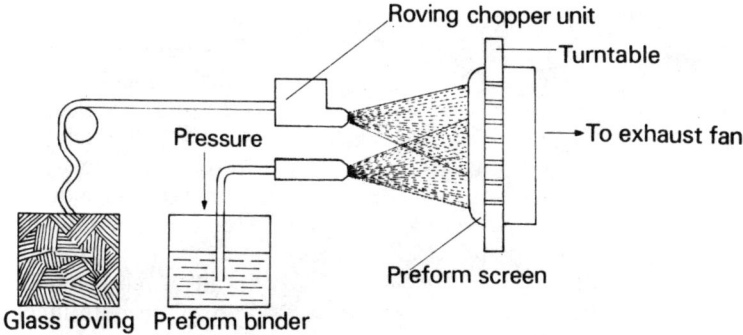

Fig. 4.10 *Diagram showing the technique of producing a preform*

separately into the mould. The glass, either in the form of tailored sheets or as a 'preform' being placed in position and a predetermined quantity of catalysed resin poured over the fibres. In the majority of cases it is usual to use a preform which ensures consistency of glass content and fibre distribution and at the same time greatly speeds the loading operation. The preform is made by laying chopped strand rovings on to a fine metal mesh of the same shape as the component to be moulded. The mesh is placed on a rotating table and the chopped strands of 5 mm to 50 mm in length are blown on to the mesh until the required thickness of reinforcement is achieved. To facilitate holding the fibres in position until the preform is completed, it is usual to incorporate a low speed exhaust fan behind, or beneath, the platen upon which the mesh screen is mounted (Figures 4.10 and 4.11). The strands are then bound with a resin soluble binder either in the form of an emulsion or a powder.

Emulsions of polyester, urea formaldehyde, polyvinyl acetate and acrylic resins are all employed as binding agents. Binders can be applied using conventional spraying equipment. The mesh screen and its covering is then transferred to a forced draught oven and stoved for approximately 3 min at 125°C. After cooling the preform can then be removed for placing in the mould or stored until required.

4.8 HOT PRESS MOULDING

The hot press moulding process bears much similarity to the cold press process in that it uses accurately aligned matched moulds which give a good finish on both sides of the component. It is used for higher volume production runs that the cold press method as cycle times are in the area of 1–3 min. Tooling, however, is necessarily more

Fig. 4.11.   Spraying a large preform. (Courtesy Owens Corning Fiberglas)

expensive and necessitates the use of highly polished and heated steel moulds and an automatic opening and closing hydraulic press in order to achieve economic production rates. The facility to control with accuracy the approach and final closing speeds of the press is also of vital importance in view of the rapid gelling resin formulations used. The pressures involved lie around 49 kgf/cm^2. Temperatures of the moulds are in the range 100–170°C. Careful consideration has to be given to the size of press selected (there are a number of units designed specifically for hot press moulding available commercially) as the size range of component to be produced will govern the platen area and the distance between the platens when in the fully open position.

The design of moulds for the hot press process (see Chapter 10) is again more critical than for the lower pressure cold process. For the longer production runs, moulds are of high grade steel, polished and hard-chrome plated with hardened cut-off edges. Shorter runs can be produced using various types of alloy for the moulds but these will be required to incorporate hardened edge inserts around the periphery of the component area.

In the moulding operation, glass reinforcement commences, as in the cold press process, with the laying up of dry glass fibre reinforcement, usually as a preform, over the male half of the mould and a measured quantity of hot curing catalysed resin poured over the preform. The press is then closed quickly with the final travel accurately controlled at a slow speed to permit complete impregnation of the reinforcement prior to the heat from the mould halves being transmitted to the embryo laminate.

As in the cold moulding process it is sometimes necessary to apply a gel coat to either or both of the mould surfaces (Section 4.7). If the gel coat can be sprayed into a hot mould a fairly short cycle can be obtained but it must be noted that unless the gel coat is allowed to cure its effect will be lost and a fibre pattern will appear on the component surface. Additionally, as pressure is used in both the cold and hot press moulding processes, even with a full gel coat some fibre pattern is usually apparent.

4.9 SMC MOULDING

Sheet moulding compounds (SMC) are one of the more recent developments in the GRP field and an advance which has contributed greatly to the use of glass reinforced plastics in automobile bodywork. The material is now considered for use in the design stage of a production body and not merely as a substitute for metal. One factor which has encouraged its use both in the U.S. and in Europe has been the introduction of low shrink and low profile materials. As discribed in Section 3.3.6 these are materials with centreline averages of less than 0.25 microns and exhibiting no long term waviness. They can be moulded to meet the many and various structural and dimensional requirements in a modern body hitherto unattainable with the earlier materials and methods of moulding and can be painted after a suitable degreasing operation. Additionally, the cost of SMC materials and rapid moulding cycles attainable show a very favourable cost/performance relationship particularly for high production run applications.

In the equipment area the application of high quality SMC parts by quantity manufacturers has led to modifications and improvements in hydraulic press designed for use with glass reinforced plastics and a number of equipment manufacturees produce presses designed specifically for use with SMC. These machines provide pressures in the range of 35–140 kgf/cm^2. Choice of pressure is determined by the type of SMC used and the shape and size of the mould. Low shrink, low profile SMC materials are relatively highly filled and these high glass types require greater pressures in order to obtain good surface quality. Additionally parts of deep draw require higher moulding pressures than generally flat parts.

As in the other press moulding processes, presses are required to have rapid advance and slow final closing characteristics in order to permit the SMC to flow in the mould before full pressure is applied and cure commences. Close control of the final closing cycle is vital as too high a closing speed can force the resin away from the glass fibre reinforcement. The actual closing rate will depend on factors such as mould temperature and gel time and the faster curing materials will naturally require higher closing speeds and more rapid build-up of final pressure. Normally, final closing speeds are in the range of 1–10 s and are determined by trial and error, first, to prevent pre-gelling in the mould if too slow or resin wash if too fast. A general specification of a typical modern SMC press is given in Section 11.16.1.

Moulding temperatures selected to give reasonably rapid curing range from about 130–165°C using the new catalyst systems developed by SMC manufacturers. It is important that the temperature profile across the mould surface should be constant as large variations in mould surface temperature can lead to premature gellation in areas which are at a higher temperature than others.

SMC materials are produced in various thicknesses and it is invariably necessary to load the mould with a number of plies. These should be cut to a size that will give between 30–40% flow in the mould. Where plies are built-up they should be laid in a pyramid shape with each additional ply somewhat smaller than the one beneath. In this way air entrapment in the mould is avoided as the air is forced outwards as the resin flows to the periphery of the mould. Correct laying of the SMC (termed the charge pattern) ensures equal flow of the material to all the cavity extremities and obviates orientation of the fibres which can result from too great a flow in the mould – a situation which may produce uneven strength in the finished component.

Curing time of an SMC moulding is partly a function of the catalyst system used and mould temperature. The thickness of the moulding will, as in other processes, have its effect. A good indication of cure time can be obtained by placing a sample of material in the mould until it gels. This gel time will indicate the maximum press closure period while cure time can be determined by checking the hardness of the laminate.

Another method for determining the cure time of a particular SMC material is to use a standard tool designed to give a thermal profile of the material throughout the curing cycle. Thermocouples in the tool cavity indicate the temperature of the laminate through heat-up to peak exotherm and subsequent cooling to mould temperature. The curing period is shown as the number of seconds taken for the material to pass through the heat-up stage to a point just beyond peak exotherm. This method provides an accurate and reproducible cure time for an SMC material before it is put into production.

Moulds used for SMC are designed to withstand considerably higher pressures than those used in other GRP press moulding processes and thus are almost universally made of a high strength steel. They are hard-chrome plated in order to produce highly finished mouldings that require a minimum of post moulding work.

Heating of SMC moulds can be by steam, oil or electricity. The former method is the most usual as it permits close control of temperature although in many shops it is not as convenient to apply as electricity.

Moulds are fitted with travel stops to ensure consistency of component thickness in the same way as other press moulding processes described and also incorporate automatic cut-off edges around the mould periphery. In some cases moulds are fitted with ejectors, which linked to the press, facilitate extraction of the component. Some SMC moulding lines now incorporate automatic loading/unloading 'automats' to speed production and relieve the operator of the task of handling the hot mouldings.

4.10 RESIN INJECTION MOULDING

This title really covers two variants of the process. The first uses a lightweight, evacuated mould and relies on a combination of gravity and atmospheric pressure for injection of the resin. The second, as the term suggests, uses a heavy rigid mould into which the resin is injected under positive pressure.

4.10.1 Gravity methods

This method in its simplest form consists of a rigid female mould and in some cases an elastic covering membrane which forms the inner, or upper mould half. Glass fibre reinforcement is laid up in the mould, if required over a previously applied gel coat, covered with the membrane which is sealed around the periphery of the mould and a vacuum drawn to remove the majority of the air from between the two. Catalysed resin of a viscosity of 1300–1400 CP from a raised container is then allowed to flow into the reinforcement. When the required quantity of resin has been fed in, the flow is stopped, the injection point plugged and the laminate allowed to cure. Evacuation produces a moulding pressure of about 14 lb/in^2 (1 kp/cm^2). Glass fibre reinforcement ratios from 25–40% can be obtained. Reinforcement can be in the form of random mat and woven roving with, in some cases, the addition of sisal mat to aid resin flow to the extremities of the mould.

Advantages of the system are the use of a lightweight mould which can be of GRP, low capital cost of moulds and equipment and suitability for large area components including sandwich construction in comparatively small numbers. Disadvantages are that mouldings cannot be guaranteed as being free from voids, long flow paths can cause problems and too high a vacuum can cause evaporation of styrene in the resin resulting in curing problems and also in the formation of voids.

From an ecological and health point of view the use of a closed mould system virtually eliminates styrene emission even when using conventional resins and thus conforms to the aims of legislation now being introduced world wide with regard to the presence of styrene in GRP workshops.

A modification of the above system has been developed by Hoechst A.G., of W. Germany. Although necessitating somewhat higher equipment costs it does not involve greater labour input and offers a method of mechanising the medium scale production of void-free parts with a mould finish on both inner and outer surfaces. The basis of the process is the use of a rigid lower mould in GRP, a suitable and slightly flexible upper (or inner mould) and the provision of an adequate vacuum channel around the periphery of the lower mould. As in other closed mould resin injection techniques the Hoechst process involves the laying up of glass fibre reinforcement before resin is introduced. After laying up, mould closure and the application of vacuum, resin is allowed to flow into the mould by gravity. Because of the low pressures involved the lower mould which in the majority of cases forms the outer surface of the component, can be of comparatively light construction. It is recommended by Hoechst that large moulds be about 10 mm in thickness and that in order to withstand attack by styrene and acetone the gel coat should be of Alpolit VU 2178.

The flexible mould half is an important part of the technique. It should be not more than 5 mm in thickness for large moulds and constructed using a chemical resistant gel coat resin such as Alpolit UP 745 and Alpolit UP 130 in a 1:1 ratio. The laminating resin recommended is of similar formulation to impart both chemical resistance and the required degree of flexibility. The laminate should be on the resin rich side with reinforcement consisting of tissue, 300 g/m^2 chopped strand mat or mats depending on thickness required, followed by woven cloth. This build-up gives a somewhat transparent laminate which allows the operator to follow the advance of the resin during its injection. Details of mould construction are given in Chapter 10.

Gel time is of major importance and it is recommended that gel time/inhibitor content/temperature graphs are prepared to ensure that correct times are achieved. Normally chopped strand mat with an insoluble binder can be used as reinforcement but continuous strand mat can be used. The choice depends on the component strength required. The lowest glass resin ratio is obtained using chopped strand mat by itself but about 40% glass content can be obtained with a 1:1 ratio of chopped strand mat and woven roving mat. The ratio rises to about 50% when using woven roving mat alone. In laying up the reinforcement it is possible to use tissue mat to eliminate fibre pattern while staples and stitches can be used to hold the mats in position.

The size of the resin container and the number of polyethylene hose feeds to the mould will depend on the size of the component. The container is placed some 2 m above the mould to give good gravitational flow. Resin inlets and vacuum outlets on the upper and lower mould halves are of 12 mm bore metal tube with the resin inlets tapered to facilitate removal of cured resin.

The pump used by Hoechst in its demonstration equipment consists of a motor, water circulation vessel, air filter and two valves for control of vacuum. A bleeder valve allows air into the system for adjustment of vacuum. Reinforced PVC pipe is used to connect the pump to a pressure pot which retains excess resin and for connecting the pump to the high vacuum and the pot to the low vacuum outlet in the periphery of the lower mould.

Operation

After the moulds have been waxed and polished, a PVA release agent is applied in the conventional manner and the reinforcement placed in position, the upper mould half is positioned and high vacuum applied to close the mould. The feed pipes to the upper half are connected and clamped some 100 mm above the inlet orifices. The mould is again evacuated to consolidate the reinforcement. Resin is then allowed to flow through the feed pipes as far as the clamps. These are then released and repositioned lower down to allow the entrapped air to rise to the surface of the resin container. The clamps are then opened and the resin allowed to flow into the mould. The clamps are again closed when about 80 mm of resin remains in the hose. Resin flow can be assisted, when necessary, by the application of low vacuum. The hoses are then cut at 80 mm above the resin level and a low vacuum applied until the resin in the mould rises to within 75 mm of the top edge of the mould. Increased vacuum is then applied for a period of about 5 min to ensure total impregnation of the reinforcement. The mould is then left until the resin colour changes indicating gelling. As the laminate commences to exotherm and a portion of the mould is released due to shrinkage, the low vacuum is released. The moulding can then be released from both mould faces by applying compressed air to the low vacuum inlets. The upper mould half can then be removed and the finished component extracted. A typical production cycle for a moulding containing 22 kg of resin using the Hoechst method would be of the order of 75 min.

Points to note when using this method are that all the air in the inlet tubes must be evacuated prior to resin injection if an unacceptable concentration of air bubbles in the moulding are to be avoided, that the gap at the 'pinch off' area is sufficient to permit acceptable filling rates and that the upper mould half has sufficient flexibility over this area to accept the resin.

The process can be used in conjunction with filling materials to produce sandwich components incorporating cores of sealed end grain balsa wood, various plastics foams such as PVC and polyurethane or more solid, strength giving cores such as plywood.

4.10.2   Pressure methods

Resin injection systems using pressure to inject the resin have gained considerable popularity and the technique is now used by a number of major automobile manufacturers, among which are Matra Simca who produced the 'Bagheera' sports car in France and reportedly, Lotus Cars in the U.K. Advantages of the process for high economy, medium run automobile body production include deep draw facilities, the capability of using heavily filled, fire retardant resins, automatic resin handling, close dimensional control, low operator skill, low porosity mouldings and the facility of using one machine to serve a number of moulds. These advantages characterised by pressure injection are additional to those of good surface finish on both sides of the moulding, and virtual elimination of styrene emission afforded by the gravity/vacuum method described.

However, in overcoming the lengthy impregnation times associated with the gravity system, the higher pressures required in the pressure injection system necessitates the use of more costly moulds and a resin supply system capable of mixing and feeding some 6 kg of resin/min under pressure is required. Thus the selection of the system to be used is again related to production requirements as regards total quantity and quality in relation to production time. If it is assumed that for the above reasons a pressure injection system is to be used, the characteristics of the component itself will determine the impregnation pressure. This in turn will be dependent upon the size of the moulding, cycle time required, viscosity of the resins specified, the ratio of reinforcement/resin, gel time limitations and lastly the cost of the moulds, which can be up to four times more expensive as similar moulds for hand or spray-up use.

Briefly, the larger the component and the shorter the cycle required, the higher the pressure needed and thus the heavier and more costly the moulds.

Moulds are basically similar in principle to those used in the low pressure system but are of much heavier construction due to the pressures involved which can range from 70–210 lb/in^2 (5–15 kgf/cm^2), can be of GRP or glass reinforced epoxy resin. These types on average have a production life of about 2000 and 4000 cycles respectively. For longer runs of up to 5000 cycles, sprayed bismuth/tin alloy moulds, polyester or epoxy backed, are suitable. Fabricated steel, where suitable, or sprayed zinc/aluminium GRP or epoxy backed give runs up to 10,000. For very extended runs, of the order of 50,000, it is economic to use electroformed nickel copper moulds suitably backed with a reinforced epoxy resin. As detailed in Section 10.8 heating elements can be incorporated during mould construction.

As with other moulding processes new moulds require special care when being put into operation. Multiple wax coatings should be applied, followed by polishing and the application of a release agent such as PVA. Some moulders use a resin which incorporates a release agent which facilitates release. This formulation in conjunction with a mould surface resin filled with up to 15% of PTFE powder can obviate the need for any surface release preparation from cycle to cycle.

Injecting the resin

Small and regular shaped components can be injected at a single central point using an inlet pipe some 6–7 mm internal diameter.

Large and complex shaped components are injected at a number of points. This reduces strain in the mould and also reduces injection time. Injection can be·simultaneous at all points or sequential depending on the flow characteristics of the mould form. After injection each point is plugged to eliminate back flow. As the resin enters the mould air is evacuated at the pinch-off area and any excess resin rises into a channel moulded in the periphery of the lower mould half. In some mould designs a rubber grommet is inserted along the pinch-off area to guide excess resin through a vent tube. When this tube fills it indicates that the mould is full and injection complete. Some of the latest resin injection equipment incorporates an automatic mould shot facility which permits the operator to work elsewhere while injection progresses.

Inserts and core materials can be incorporated in the moulding. These are attached to the mould during the lay-up operation. As in the low pressure process various core materials such as balsa or rigid/semi-rigid foam can be incorporated to form a composite structure. The core material is interlayered between the glass fibre reinforcement prior to injection.

4.11 COMPOSITE MOULDINGS

The term composite moulding covers a number of techniques of combining GRP laminates with other materials to achieve an optimum stiffness/weight ratio. As with a number of other methods of moulding GRP the early techniques originated in the aircraft industry where the properties of composite construction showed to particular advantage and where cost was of less consequence than in other areas. While the methods reviewed briefly here complete the list of basic moulding processes, it cannot be claimed that they are used to any great extent in automobile body construction largely because of cost considerations. However, for certain components and particularly for specialist production a double-skinned or sandwich structure incorporating a lightweight core can provide notable stiffness with a minimum weight penalty. Designers are fully aware of the advantages of such a structure and this form of sandwich construction is now widely used in the injection moulding of thermoplastics. Development are also continuing in an attempt to reduce the labour content in the production of double skinned GRP mouldings.

4.11.1 Hand and spray lay-up with lightweight core

The advantage of using two thin skins of GRP with a lightweight core rather than a single thick skin stems from the fact that the former method of construction is far more rigid weight-for-weight than the latter. Basically there are two methods of producing a lightweight core moulding using the hand or spray lay-up technique. In the first a thin laminate is laid-up in the mould in the usual way and before the resin has gelled a layer of rigid plastic foam, end grain balsa wood blocks or resin impregnated paper honeycomb is placed in position and pressed lightly into the wet upper layer of reinforcement. When the laminate has gelled, the inner skin of the moulding is laid up, again in the conventional manner but with considerable care especially when rolling in order not to push the reinforcement through the honeycomb when this type of core material is used.

4.11.2 In-situ foamed core

The second method of producing a rigidized double skin moulding necessitates the use of two moulds, one for the outer and one for the inner skin. When both skins have cured they are joined around their periphery and the interspace is injected with one of the in situ foaming formulations now available. This method produces a rigid lightweight moulding with a mould finish on both inner and outer surfaces. Apart from the cost of an additional mould there are limitations regarding the type of moulding that can be handled successfully. It will be appreciated that the gassing agent used in the foam formulation produces a considerable pressure as foaming proceeds in the closed moulding. For this reason when foaming large relatively flat mouldings, such as a bonnet or boot lid, it is necessary to carry out the operation with both skins held either in their moulds or in a specially built foaming fixture (Figures 4.12 and 4.13). Failure to hold the skins over their entire surface area will result in distortion caused by the internal pressure. Open ended mouldings are less prone to this effect and are simpler to rigidize as the liquid foam formulation can literally be poured in without the need for even the simplest of injection equipment.

4.12 RESERVOIR PROCESSES

In this process[6] a closable mould is laid up with dry reinforcement and closed on to a resin impregnated open cell foam core which serves as the resin reservoir (Figure 4.14). As the mould is closed sufficient resin is squeezed out of the foam to impregnate the two layers of reinforcement. The system offers a number of advantages for the production of a sandwich laminate. It involves the impregnation of

Fig. 4.12.   Foaming jig for the rear door of a commercial vehicle. (Courtesy Pressed Steel Fisher BL Components Ltd)

only a single layer of material irrespective of the final moulding required, while the soft open cell foam readily drapes to the shape of all but the most complex of moulds.

During mould closing the foam/reinforcement passes through the sequence shown in Figure 4.15. First, the foam core presses the reinforcement against the surfaces of the mould. On further closing most of the air in the foam is squeezed out and escapes through the still dry glass fibre. Final closing forms and compacts the reinforcement to the mould form and, as the remainder of the air is removed, resin is squeezed out to impregnate the glass fibre inner and outer skins.

The process offers considerable scope for variation in the type and thickness of both reinforcement and foam while different thicknesses and specific gravities can be obtained using the same materials by simply varying the degree of compression. Moulding pressure is an important factor and a system of stops is always used. The pressure required is that sufficient to compress the foam core to a predetermined point. Normally the pressure is of the order of $1–2\,\mathrm{kgf/cm^2}$ for a large generally flat moulding in high viscosity or a highly filled resin.

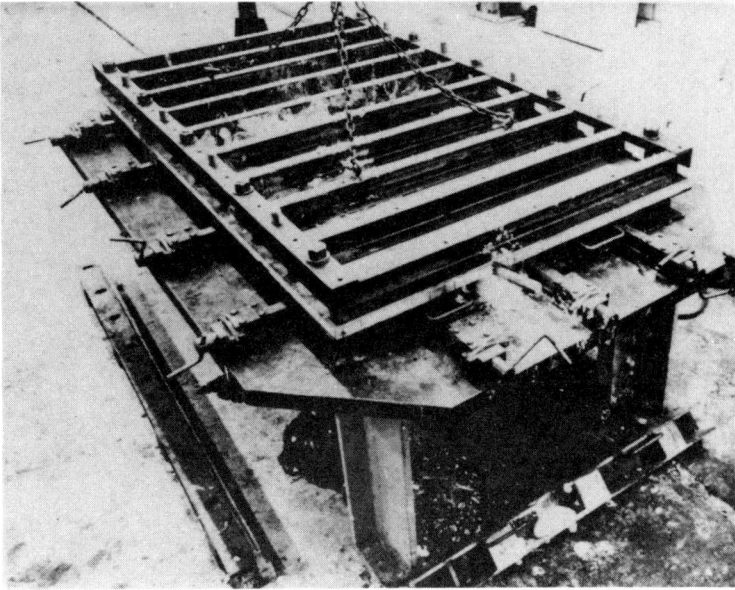

Fig. 4.13. *The foaming jig closed for the injection of the foaming resin.* (*Courtesy Pressed Steel Fisher BL Components Ltd*)

Moulding can be carried out at temperatures ranging from ambient to 180°C giving cycle times of from seconds to hours. Given these conditions it will be appreciated that the resin system used is of considerable importance and particular attention must be paid to viscosity both at room temperature and with variation with time at moulding temperatures. Viscosity must be correct for initial impregnation of the foam and for retention of the resin prior to moulding. Again if viscosity is too low at moulding temperature premature impregnation of the reinforcement can occur.

The most important processing advantages are that moulding pressures are low and that moulding at shop temperature can be carried out using GRP moulds. Higher temperatures used in conjunction with matched metal moulds permit moulding on faster cycles and without the need for a press.

4.13 DUCT WINDING

Where limited numbers of hollow components of complex shape such as ventilating ducts are required they can be produced economically

Table 4.1 CHARACTERISTICS OF THE MAJOR PROCESSES (COURTESY COMPOSITES LTD.)

	Contact moulding		Matched die moulding	
	Hand lay-up spray-up	Dough moulding	Mat preform pre-preg.	Cold press
Maximum part size determined by	Mould size, transport of part	Press rating and size	Press rating and size	Press rating and size
Shape and styling limitations	None	Mouldability	Mouldability	Mouldability
Volume of production category	Low to medium	High	High	Medium
Number of finished surfaces and quality	One Excellent	All Excellent	Two Very good	Two Good
Typical glass content by weight	20–35%	10–35%	25–45%	20–40%
Strength category	Medium	Low/medium	Medium/high	Medium
Strength orientation	Random (except fabric types)	Random throughout moulding	Random	Random
Resin rich, corrosion resistant surfaces	Yes, by gel-coat and/or surface mat	No	Yes, by gel-coat and/or overlay mat*	Yes, by gel-coat and/or surface overlay mat
Practical thickness range (mm)	0.76–25.4	1.5–25.4	0.76–6.3	1.5–12.7
Common moulding tolerance on thickness (mm)**	±0.5	±0.05	±0.21	±0.5
Local thickness increase	As desired	As desired	Usually 25:1	Usually 25:1
Metal inserts and/or edge stiffeners	Possible	Possible	Possible	Possible

Table 4.1 continued

	Contact moulding		Matched die moulding	
	Hand lay-up spray-up	Dough moulding	Mat preform pre-preg.	Cold press
Built-in cores	Possible	Possible	Possible	Possible
Minimum radii for ease of moulding (mm)	13	0.80	0.32	0.50
Undercuts	Yes	Yes	No	No
Minimum recommended draw angle (degrees)	2	1	1	2
Holes moulded in to avoid material waste	Yes (large)	Yes	Yes	No
Trim in mould	Yes (rough trim)	Yes (except fine flash)	Yes (except fine flash)	No

* Except pre-preg.
** Dependent upon laminate thickness

Inner glass fibre skin

Male mould

Female mould

Outer glass fibre skin   Resin impregnated foam core

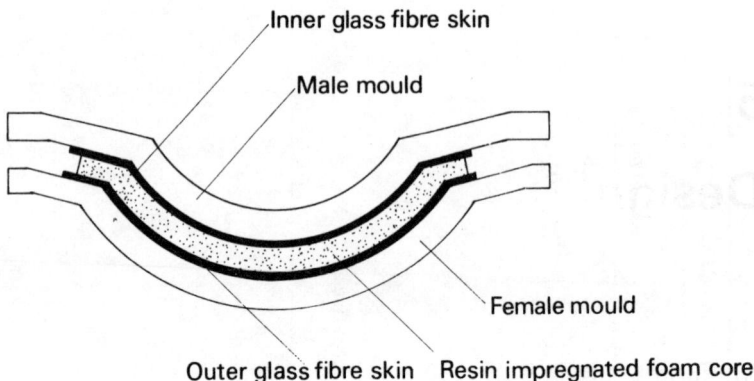

Fig. 4.14.   *Diagram illustrating the principle of Reservoir moulding*

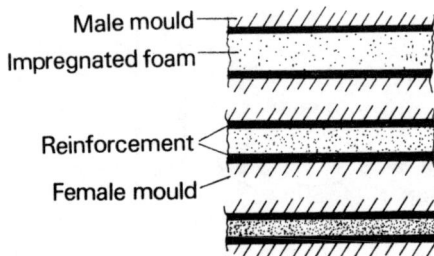

Male mould
Impregnated foam

Reinforcement
Female mould

Fig. 4.15.   *Sequence followed during closing of the mould to squeeze the resin from the porous core into the inner and outer fibre skins*

by winding strips of resin-impregnated woven glass cloth or tape around a former which is later removed. Formers can be of two types depending on the shape of the component. Simple forms that permit the former to be collapsed to remove the moulding can be of wood or for longer runs, aluminium or castable alloy.

Complex shaped components are wound on a hollow plaster former which can be broken after the component has attained full cure. Where runs of these components are extended it is usual practice to build a mould for producing the plaster formers. In cases where it is difficult to make a hollow plaster former hollow it can be cast around a steel spring or bunch of cords. When full cure has occurred pulling on the end of the spring or cord will facilitate breaking out the plaster. Excellent interior surface finish can be obtained by liberal coatings of wax over the former prior to gel coating and lay-up and by winding polyethylene tape around the finished duct prior to curing.

5
Design

5.1 BASIC CRITERIA: HAND OR SPRAY LAY-UP

As highlighted in earlier chapters the characteristics of GRP permit the designer very wide freedom of component form within the physical limitations of the material. However, in a car body in particular, styling and functional requirements can conflict with each other and thus the freedom offered the designer by a moulding process as compared with conventional fabrication in metal is often curtailed.

Styling plays a large part in a model's appeal to the buyer and is thus of major importance in a new design. Traditionally the body lines of the popular car have been a compromise between the desire on the part of the stylist to produce a model having the maximum visual appeal and the constraints placed upon him by the economics of quantity production in steel. In the case of the specialist body builder the economic constraints are less rigid. Production can be counted in single figures rather than in thousands and whilst labour content is very much higher tooling costs for hand or spray-up are considerably lower than those involved in the mass production area.

Whilst it is not within the scope of this book to discuss the many and complex factors which have to be taken into account in the design of a complete body, even for small scale production, some of the more basic points of design when working in GRP are covered briefly.

In view of the rapid increase in the application of sheet moulding compounds (SMC) for quantity production basic design factors are divided first into those applicable to moulding processes other than matched tooling and secondly, the moulding of the newer sheet compounds. Whilst many of the basic rules apply to both areas, to a greater or lesser extent, depending on the moulding method, the latter has now become very much the province of the production engineer where body design involves complex problems such as stressing, compatibility with mechanical components, production costs and labour relations. This particular area is, however, outside the scope of this book.

For the specialist and amateur bodybuilder and for prototyping applications, body shells and large body panels will in all probability be produced by hand or spray-up methods on the grounds of mould cost and in the case of prototypes in the need for design changes. The choice between the two methods will be governed by the availability of equipment, speed of production required and the size and quality of the labour force on hand.

From a design point of view both processes give similar freedom with regard to the inclusion of inserts and panel curvature and form and very similar results on a thickness/cost basis. Bending strength/cost will be virtually the same for both hand and spray lay-up but stiffness/cost will be somewhat higher in the case of a spray-up moulding of similar form and material content. Table 5.1 gives the physical properties and cost/performance comparison for various reinforcements laid up by the hand method and for rovings applied by spray techniques.

Characteristics common to both hand and spray lay-up can be itemised as follows:

(1) To avoid poor fibre distribution and the possible inclusion of air bubbles, inside radii on mouldings should be 6 mm minimum.

(2) Large moulded-in holes can be incorporated.

(3) Undercuts can be moulded providing the mould is split: (4) Both hand and spray-up methods permit trimming of the moulding whilst it is in the mould.

(5) Minimum draft recommended in both methods is 2°.

5.1.1 Component thickness

The minimum practical thickness of a moulding is generally accepted as being about 0.75 mm on hand lay-up and about 1.5 mm on spray-up work. There is virtually no upper limit on thickness but note should be taken of the limitations on producing wet-on-wet lay-ups as detailed in Section 4.2.3 and also of the danger of mould damage through excessive heat generated by the exotherm in thick mouldings. Wall thickness tolerance is largely dependent upon the skill of the operator and can be in the range of $+1.20$ mm to -0.50 mm on hand lay-up and $+0.5$ mm to -0.5 mm on spray-up work.

Both methods permit the use of metal inserts, metal edge stiffening, the incorporation of integral bosses and ribs and both hand and spray-up methods can be used in conjunction with a gel coat and, if required, surfacing mat. The important advantage of GRP con-

Table 5.1 COST/PERFORMANCE COMPARISON FOR MATERIALS (COURTESY OWENS CORNING FIBERGLAS)

Material	Glass content by weight (%)	Material cost new pence/kg	Bend strength (BS)		Stiffness (ST)		Tensile	
			Thickness (mm)	Material cost (100 mm²/ new pence)	Thickness (mm)	Material cost (100 mm²/ new pence)	Thickness (mm)	Material cost (100 mm²/ new pence)
Lay-up mat	35	33	0.28	13	6.8	31	1.7	17
Lay-up mat M 700 fine								
Spray-up rovings	25	26	0.33	12	7.4	26	2.1	15
Lay-up cloth	45	55	2.2	20	5.7	47	0.8	15
Lay-up woven roving	50	35	2.1	12	5.4	29	0.6	7
Tabmat-2415 lay-up								
Aluminium	—	55	4.0	60	3.2	46	2.0	61
Steel	—	9	2.5	15	2.1	12	0.8	10
Stainless steel	—	66	2.3	117	2.2	110	0.7	81

Table 5.2 HAND LAY-UP. AVERAGE THICKNESS IN MM PER NUMBER OF PLIES (COURTESY OWENS CORNING FIBREGLASS)

Laminate	Number of plies														
	1	2	3	4	5	6	7	8	9	10	11	12	13	14	15
2 oz mat (57 g)	1.5	2.8	4.5	6.0	7.5	9.5	10.75	12.5	13.5	15.5	17.0	18.75	20.25	22.0	23.0
24 oz woven roving (680 g)	0.93	1.7	2.5	3.6	4.7	6.0	6.5	7.6	8.5	9.5	10.5	11.5	12.4	13.3	14.3
10 oz cloth (280 g)	0.4	0.78	1.2	1.6	2.0	2.4	2.8	3.2	3.6	4.0	4.4	5.0	5.3	5.7	6.2
Fabmat 2415	2.0	3.8	5.6	7.6	9.5	10.5	12.7	14.5	16.5	18.4	20.3	22.2	24.1	32.0	34.0

struction in that local thickening can be used where necessary by both methods of lay-up.

5.1.2   Determining thickness/stiffness

The thickness of the component is the single most important factor to be determined in a moulding as it effects directly the quantity of reinforcement and resin to be used and thus the cost of the part. As already mentioned in Section 1.1 and in some detail in Section 5.5, the shape of the component will also have an important part to play in determining stiffness. The two factors, thickness and shape must, therefore, be considered together.

Table 5.1 can be used to compare hand and spray-up laminates with each other and with some widely used conventional materials on a cost/thickness basis for equal performance with regard to specific properties. The chart is applied in the following manner: If for example it is assumed that the two main criteria of a moulding are bending strength (BS) and stiffness (ST), the selection of these will be between woven roving reinforced hand lay-up, spray-up and aluminium. From Table 5.1 the comparative cost in new pence versus performance in terms of bend strength and stiffness can be seen. For example, the costs in new pence per 100 mm^2 of (1) a hand lay-up in woven roving, (2) a spray-up and (3) aluminium sheet of equal bend strength are 12, 12, and 60p respectively. Thicknesses will be 2.1, 0.35 and 4 mm respectively. Similarly, the cost in new pence per 100 mm^2 of the same materials relative to stiffness will be 0.6, 2.1, and 2.0 p. Thicknesses will be 5.4, 7.4 and 3.2 mm respectively. Therefore, as the spray-up roving is more economic to use on both counts, it is the material selected. However, selection is based solely on material cost and for the complete breakdown it is necessary to consider what production rates are required. Table 5.2 shows the number of plies of laminate in hand lay-up for a given average thickness of laminate. The designed thickness can be obtained (1) by comparison where the thickness of a part in another material is known: (2) theoretically, where the specific strength requirements are known. Here the calculations can be made with the aid of standard Strength of Materials tables: (3) thickness can be determined empirically using past experience of successful applications.

5.1.3   Comparative method

If the thickness of a part in another material is known, Table 5.2 can be used to calculate the required thickness of the moulding in GRP to give comparable stiffness using the following formula:

Assume required thickness of GRP = GRP_{Th} and thickness of conventional material = Con_{Th}, then

$$GRP_{Th} = Con_{Th} \times \frac{\text{Factor for GRP}}{\text{Factor for conventional material}}$$

Example 1

A part in aluminium is 1.27 mm in thickness. Then the thickness of a GRP moulding by spray-up of comparable stiffness will be:-

$$GRP_{Th} = 1.27 \times \frac{7.30}{3.30} = 2.8 \text{ mm}$$

Thus it can be calculated quite simply that a GRP moulding 2.8 mm thick produced by spray-up will be equal in stiffness to an aluminium part 1.27 mm thick.

Example 2

A part in stainless steel is 0.051 mm thick. Then the thickness of a hand lay-up moulding in 1000 cloth of comparable stiffness will be:-

$$GRP_{Th} = 0.051 \times \frac{5.6}{2.2} = 0.128 \text{ mm}$$

In using these simple calculations the thickness limitations of GRP must be taken into account. In Example 1 a moulding 2.8 mm in thickness is practical for both hand and spray-up processes. However, in Example 2, where the calculated thickness is only 0.128 mm, this cannot be obtained by either laminating method and therefore the minimum practical thickness of moulding must be used.

5.1.4 Theoretical method

As mentioned in the previous Section, theoretical design can be applied when the stress condition under which the moulding will operate is known. So far in this chapter we have only considered single reinforcement but as shown in Table 5.3 it is possible to use

Table 5.3 THICKNESS, PERCENTAGE OF GLASS CONTENT, TENSILE AND FLEXURAL STRENGTH OF LAMINATES WITH VARIOUS REINFORCEMENT COMPOSITIONS. (COURTESY OWENS CORNING FIBERGLAS)

Laminate	Plies	Material	Glass content (%)	Tensile strength (kgf/cm²)	Flexural strength (kgf/cm²)
1	2	2 oz (56.7 g) mat	28.2	983	1638
2	1 1	1000 cloth 2 oz (56.7 g) mat	28.4	735	1298
3	1 2	1000 cloth 1.5 oz (42.7 g) mat	32.0	969	1547
4	1 2	1000 cloth 2 oz (56.7 g) mat	30.1	1046	1498
5	1 3	1000 cloth 2 oz (56.7 g) mat	32.0	924	1312
6	1 1 1	1000 cloth 1.5 oz (42.7 g) mat 1000 cloth	24.9	636	2632
7	1 2 1	1000 cloth 1.5 oz (42.7 g) mat 1000 cloth	22.6	735	1988
8	1 3	1.5 oz (42.7 g) mat 2415 Fabmat	39.2	2121	2996
9	1 1 1	1000 cloth 24 oz (680 g) mat 1000 cloth	42.5	1295	1995

Table 5.3 continued

Laminate	Plies	Material	Glass content (%)	Tensile strength (kgf/cm²)	Flexural strength (kgf/cm²)
10	1	1000 cloth	38.3	1193	1599
	1	1.5 oz (42.7 g) mat			
	1	24 oz (680 g) WR			
11	2	24 oz (680 g) WR	52.7	2726	3143
12	1	24 oz (680 g) WR	53.2	2030	3213
	1	1.5 oz (42.7 g) mat			
	1	24 oz (680 g) WR			
13	1	2 oz (56.7 g) mat	36.0	805	1610
	1	24 oz (680 g) WR			
	1	1000 cloth			
14	1	1.5 oz (42.7 g) mat	47.0	1554	2926
	2	24 oz (680 g) WR			
15	1	1.5 oz (42.7 g) mat	47.9	1743	2198
	1	24 oz (680 g) WR			
	1	1.5 oz (42.7 g) mat			
	1	1000 cloth			

combinations of different reinforcement such as mat or spray-up roving combined with cloth and/or woven rovings to obtain specific properties. Table 5.3, shows a number of properties that can be obtained by a combination of materials. The table is applicable only to laminate produced by hand lay-up as woven roving reinforcement cannot be applied by spray methods. However, the properties shown also apply to any combination laminate which utilises spray-up rovings in place of random mat.

5.1.5   Empirical method

During the past years a considerable quantity of performance data has been tabulated regarding specific types of moulding produced by both hand and spray-up methods which can be consulted prior to solving a design problem. Table 5.2 shows the designer the quantity of reinforcement required in the case of a spray-up moulding to obtain a given thickness and in the case of hand lay-up, the number of plies required. Where more than one type of reinforcement is used, the empirical method shows to advantage as data is not normally available for all the combinations of reinforcement. In addition design loads are not accurately tabulated.

Example 1

To determine the thickness of a laminate consisting of one ply of 280 g (10 oz) cloth, one ply of 57 g (2 oz) mat and five plies of 680 g (24 oz) roving:

$$
\begin{aligned}
\text{1 ply of cloth} &= 0.40 \text{ mm} \\
\text{1 ply of 57 g (2 oz) mat} &= 1.5 \quad \text{mm} \\
\text{5 plies of roving} &= 4.7 \quad \text{mm} \\
&= 6.60 \text{ mm total}
\end{aligned}
$$

The final selection must take into account the shape and complexity of the moulding, the difficulties of moulding, the conditions under which it will operate in service, its size, the finish required and the numbers off required. The latter three factors are obviously of major importance in a body moulding which, while looking good, must withstand weather conditions over a lengthy period without surface crazing due to undue flexibility.

5.2   DESIGNING-IN SAFETY FACTORS

In the design of a moulding, consideration must be given to an adequate safety factor based on an assessment of predicted load and stress that will be involved in service. Minimum factors based on various types of loading can be itemised as follows:

(1) Static or short term loads, safety factor 2
(2) Static long-term loads, safety factor 4
(3) Variable loads, safety factor 4
(4) Repetitive loads, safety factor 5
(6) Fatigue or load reversal, safety factor 5
(6) Impact loads, safety factor 10.

It will be appreciated that these safety factors are considerably higher than those used when designing in metals because the strength of a laminate depends very greatly on the skill of the operator and the conditions under which the moulding is produced. Therefore it is not possible to be so specific when dealing with GRP and in particular moulding methods other than SMC. Properties should be viewed in ranges rather than as specific values.

5.3   PRACTICAL POINTS OF DESIGN: JOINS

Hand and spray lay-up processes simplify part design by permitting large structures to be produced in one-piece. However, parts must not be designed in such a way that operators have a problem in reaching central areas of large moulds in order to ensure adequate impregnation and rolling out of the reinforcement. The positioning of stiffener ribs and inserts is also of importance in the design of a large moulding, not only from the aspect of access but also from the economic point of view. A difficult access area in a large mould may lead to poor lamination and will certainly result in longer moulding times than necessary.

As described in Section 10.5, large moulds, where possible, should be pivoted to facilitate operator access. The advatnage is that it is usually easier to work on a horizontal, or near horizontal surface than a vertical one. Where problems due to size and access factors preclude the use of a one piece moulding it is possible with careful design, to join mouldings prior to final assembly.

In addition to simplifying handling, another advantage that can accrue from including a join, if correctly placed, is that the thickening involved serves to stiffen the moulding over the area (Figure 5.1). Where it is feasible to butt two mouldings, such as at a styling line, they can be produced with a returned edge and joined with adhesive

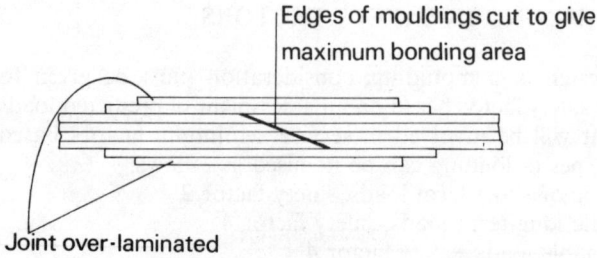

Edges of mouldings cut to give maximum bonding area

Joint over-laminated

Fig. 5.1.   *Principle of joining two mouldings*

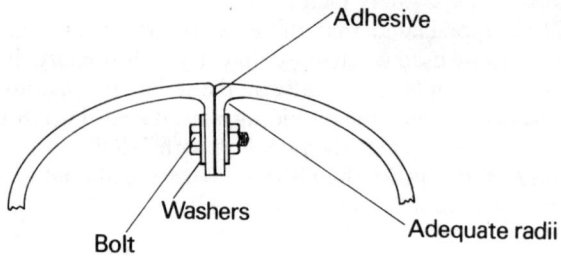

Adhesive

Washers

Bolt

Adequate radii

Fig. 5.2.   *Joining mouldings at a styling line*

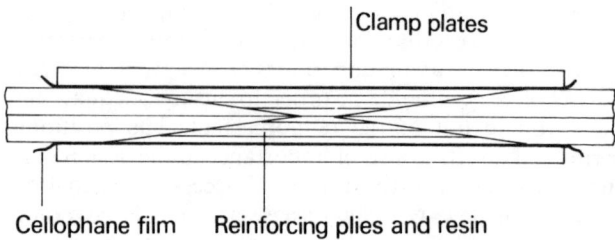

Clamp plates

Cellophane film   Reinforcing plies and resin

Fig. 5.3.   *Joining two mouldings with a flush surface*

and/or bolts of self-tapping screws. This treatment lends even greater rigidity to the assembly (Figure 5.2). Where it is necessary or desirable to join mouldings but maintain a flush surface the edges of the mouldings are chamfered at an obtuse angle and the join made between two rigid clamping plates faced with cellophane to ensure a clean parting surface (Figure 5.3).

In applications where a raised styling line is required or two mouldings join at right angles the joint can be made to serve as a visual feature and can be covered by a snap-on channel to cover the join (Figure 5.4). Depending on the application, the joint can be made

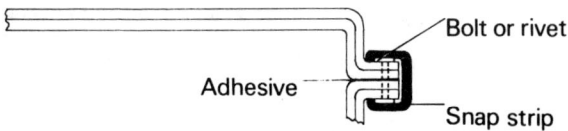

Fig. 5.4.   *Joining two mouldings to serve as a visual feature*

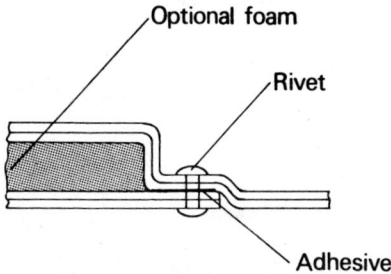

Fig. 5.5.   *Joining mouldings where double skins are involved*

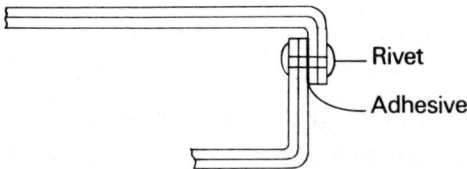

Fig. 5.6.   *Lap joint with rivet and adhesive*

permanent using an adhesive and rivets or in the case of a body moulding which may have to be replaced the joint can be filled with a thin rubber strip and bolts or screws used in place of rivets.

Another useful method of joining panels, such as floor sections, where double skinning is used to give rigidity is shown in Figure 5.5. Again, the joint can be made permanent using an adhesive and/or rivets, or the panels can be replaceable by the use of bolts only. In many cases additional rigidity, insulation and sound deadening can be obtained in permanent joints by filling the interspace with self-foaming resin mixture which sets to a rigid lightweight foam (see Section 4.11). Lapped sections can be joined permanently using an adhesive and/or rivets, with the joint covered by a trim strip where necessary (Figure 5.6).

Removable and hinged access panels not requiring a rain gutter can be set flush with the surrounding bodywork by the use of a moulded

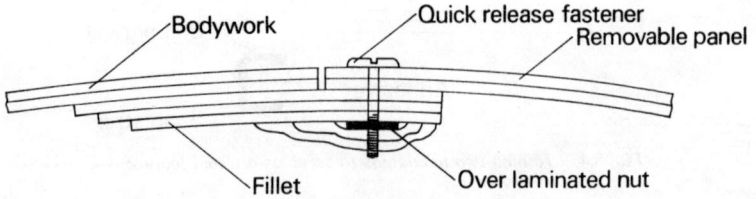

Fig. 5.7. One method of producing a removable panel

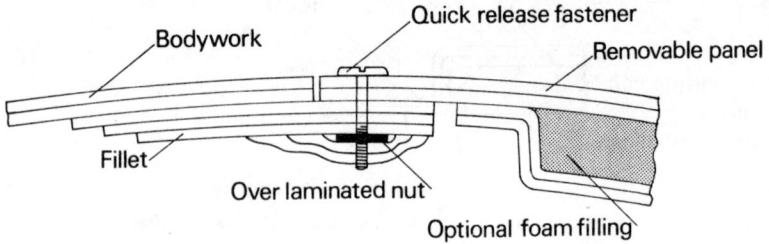

Fig. 5.8. A removable panel with double skin and foam core

fillet on which the panel is seated (Figure 5.7). The fillet can be produced on one-off bodywork by cutting out the panel with a jig-saw and setting the panel back into position with the upper surfaces flush and held temporarily with adhesive strips. With the underside of the panel liberally coated with release agent, successive layers of reinforcement can be built up around the periphery for a width adequate to accept the hinges, screws or quick-release catches for locking the panel. As a general rule where threaded fasteners are used they should always be inserted perpendicular to the laminate surface and should be laid out with edge and side distances equal to at least $2\frac{1}{2}$ times the diameter of the fastener. Spacing should exceed three times the fastener diameter. When the laminate has cured, the panel is removed and the inner edges of the fillet trimmed and smoothed. Thus when the panel is replaced its outer surface is perfectly flush with the surrounding bodywork. This system is suitable for small panels which have an inherent stiffness. Where larger panels are to be moulded some form of diaphragm stiffening (Section 5.5) or double skinning is necessary (Section 4.12, Figure 5.8).

5.4 STIFFENING METHODS: RIBS

Because GRP laminates are inherently more flexible than metal the provision of stiffening either by means of curvature in the moulding or

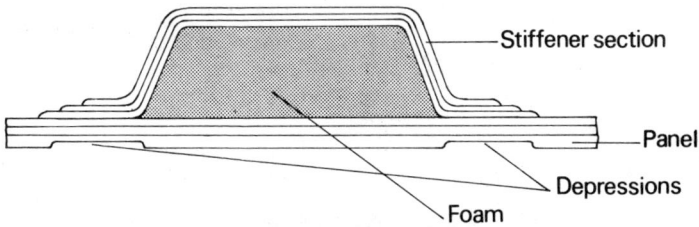

Fig. 5.9. Method of using foam filled section to provide diaphragm stiffness in a large panel

by the addition of ribs is a major design consideration. Obviously if a moulding can be designed to provide stiffness by its form alone it is more economic both from the point of view of weight of material used and labour expended rather than by providing additional stiffening either during the moulding operation or as a post cure operation.

Stiffening methods vary widely depending on the size of the moulding, its shape and the degree of rigidity required. Any form of stiffness, however, should aim at spreading loads over as wide an area as possible. One of the most convenient and widely used methods involves the use of a lightweight core of rigid or semi-rigid form overlaminated with reinforcement. The foam acts purely as a former with the hollow section laminate providing the necessary rigidity to the moulding. The operation can be carried out whilst the moulding is still in the mould and before full cure has taken place, or can be effected after cure as a secondary operation. Choice of method will depend largely upon the production cycle of the main mould and the convenience with regard to shop layouts. One factor of major importance to be considered when laying-up reinforcing stiffness before the main laminate is fully cured is that in most cases cure shrinkage of the stiffeners will cause a slight depression in the outer surface of the panel. Thus where visual surfaces are being handled the stiffening should not be applied until the moulding is fully cured. A simple foam cored stiffener is shown in section in Figure 5.9. The diagram also depicts the effect of stiffener shrinkage causing a depression on the panel surface.

The majority of panels requiring additional stiffening will be curved, albeit possibly only slightly, but enough to demand that the core material of the stiffener be sufficiently flexible so that it can conform to the inner contour of the main laminate. This requirement is met by the natural flexibility of most semi-rigid foams and thin-wall plastics tube which also forms a suitable former for the stiffener. Rigid section such as light metal, Balsa, plastics and cardboard folded into a tapered 'top-hat' section, can be partially cut through to permit them

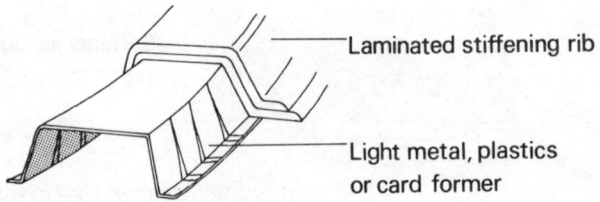

Fig. 5.10.   *Top hat section used as a former for over-laminating a stiffener*

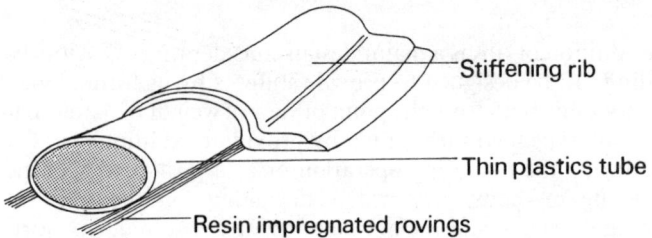

Fig. 5.11.   *The use of a thin deformable tube as a former*

to conform to the contours of the panel (Figure 5.10). Various rib former materials and sections are now available from suppliers of laminating equipment.

The operation of moulding the stiffening rib consists of bedding it down against the moulding. If the moulding is still in the mould this can be done just prior to gelling or if working on a fully cured moulding by fixing the former into position with a light application of fast curing resin. Proprietary adhesives should be avoided as in some cases the formulation reacts with the resin to inhibit full cure. The over-laminating operation can be carried out by either hand or spray-up methods but care should be taken to ensure that there is an adequate area of laminate in contact with the panel at each side of the stiffener. Some other types of former for producing stiffening ribs are shown diagrammatically in Figures 5.11 and 5.12. It will be noted that, as in general laminating practice, corners are pre-filled with bundles of resin impregnated rovings to avoid the formation of air voids and improve adhesion (Section 4.2.3).

5.5   THREE-DIMENSIONAL DESIGN

As will be appreciated from the foregoing the task of the GRP body designer is to produce a visually acceptable design concept which

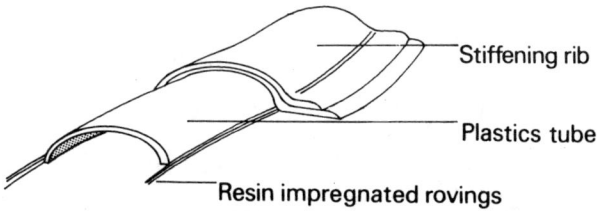

Fig. 5.12. *Half-tube used as a former*

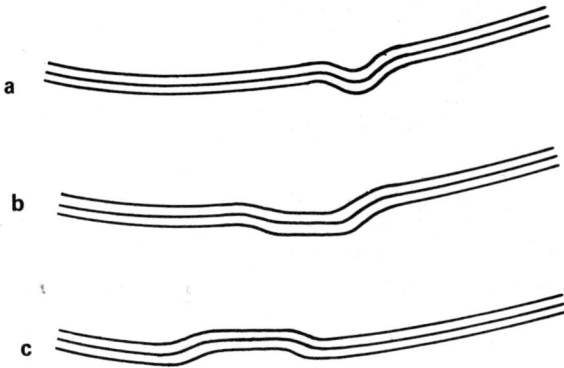

Fig. 5.13. *Exterior styling lines serve to give rigidity*

fulfils the functional demands of the vehicle, the form compensating for the material's poor stiffness. With the freedom provided by a moulding technique the most logical and, in practice, successful method is to build-in the maximum contour and where necessary to supplement the stiffness factor obtained by the use of multi-skin construction. The latter technique not only provides a means of obtaining improved rigidity but also overcomes stress problems which could only be solved in a single-skin construction by accepting a weight penalty.

The body shell lends itself to considerable design ingenuity in respect of double curvature areas which give inherent rigidity. Exterior styling lines through the body length should be incorporated in designs which lend themselves to such treatment (Figures 5.13 (a), (b), and (c)). In some cases the body form is such that the upper and lower portions, including the body floor, can be moulded separately facilitating lay-up and handling operations and on assembly providing a longitudinal flanged area which gives in-built rigidity to the body sides. The Chevrolet Corvette, one of the earliest production bodies to be produced in appreciable quantity, utilised this technique.

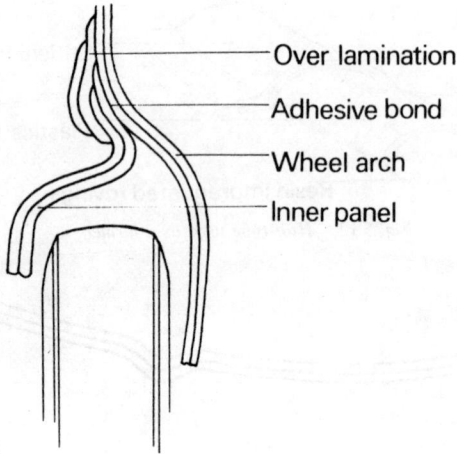

Fig. 5.14. *Diagrammatic illustration of a wheel arch stiffening*

In the early models the join formed a styling feature but in later production although the multi-part moulding technique was retained assembly was made solely with an adhesive. The external coach joint was modified to one of the reinforced channel type and the joint filled so that it was invisible on the finished body.

Wing to body areas offer opportunities for obtaining stiffness by incorporating comparatively modest concave curvature where the two flow together. Double skinning of wheel arches also adds considerable rigidity whilst serving to protect the body itself from abrasion from road debris (Figure 5.14). As most designs are such that the inner wing panel must be moulded separately and over laminated into position, usually while the body shell is still only partially cured in the mould, the laminating flanges again add to stiffness in the assembly.

Roof panels tend to present a rigidity problem as on all but the smallest of coupes they represent the largest generally unsupported area on the body. Here space limitation on headroom is a critical factor which will preclude the use of deep ribbing. Additionally, being a large visual panel any imperfections on the surface are immediately apparent. Although a double-skinned, foam-filled panel can overcome these objections, most designers specify a single skin. This requires a slight curvature coupled with a laminate thickness just sufficient to withstand panting at high speed and deflection during washing and polishing operations. It can be observed that pressed steel roofs usually can be deflected with quite a modest hand pressure if applied in the centre — a condition apparently acceptable in the

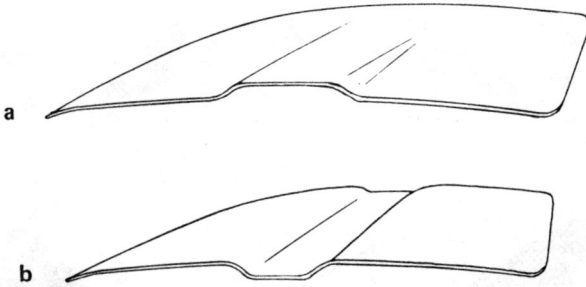

Fig. 5.15.   *Bonnet treatment to give diaphragm stiffness*

mass production car industry and one which under normal circumstances causes no problems of panting or paint cracking. Citroen in France were possibly the first quantity manufacturers to fit a GRP roof on its DS range. This large panel of modest curvature, was produced using matched metal press tooling and thus had a high glass-to-resin ratio permitting a somewhat thinner section to be used than would be practical in a hand or spray lay-up.

The incorporation of contours in generally flat areas such as bonnet, door and boot lids is still in keeping with today's styling, although the trend, in the case of the more sophisticated specialist body designer, has been away from the extreme fussiness of line so common a few years ago.

As shown diagrammatically in Figures 5.15 (a) and (b), acceptable rigidity can be obtained in even large bonnet lids by the tasteful use of raised or slightly indented styling features. The use of a fore-to-aft feature will have the effect of stiffening the panel against bending in a transverse direction but will provide little or no rigidity longitudinally or resistance to a twisting force. As it is impractical, from a visual point of view, to incorporate any form of indentations across the panel, the only solution open to the designer is to apply reinforcement beneath the area in the shape of wide but shallow ribs (Figure 5.10). However, as described, the operation would almost certainly have to be carried out after full cure to avoid the dimpling effect which would otherwise be visible on the upper surface of the panel due to shrinkage of the rib assembly.

A more acceptable and efficient solution is to use double skinning to obtain the required degree of rigidity (Figures 5.16, 5.17 and 5.18). Where the interspace between the inner and outer panels can be filled with a rigid foam it is possible to reduce the thickness of both skins in the interests of weight saving. It should be noted that the same conditions apply when joining panels around their periphery as are

Fig. 5.16. *Outer panel of a double skinned bonnet. (Courtesy Pressed Steel Fisher BL Components Ltd)*

Fig. 5.17. *The inner panel of a double skinned bonnet. (Courtesy Pressed Steel Fisher BL Components Ltd)*

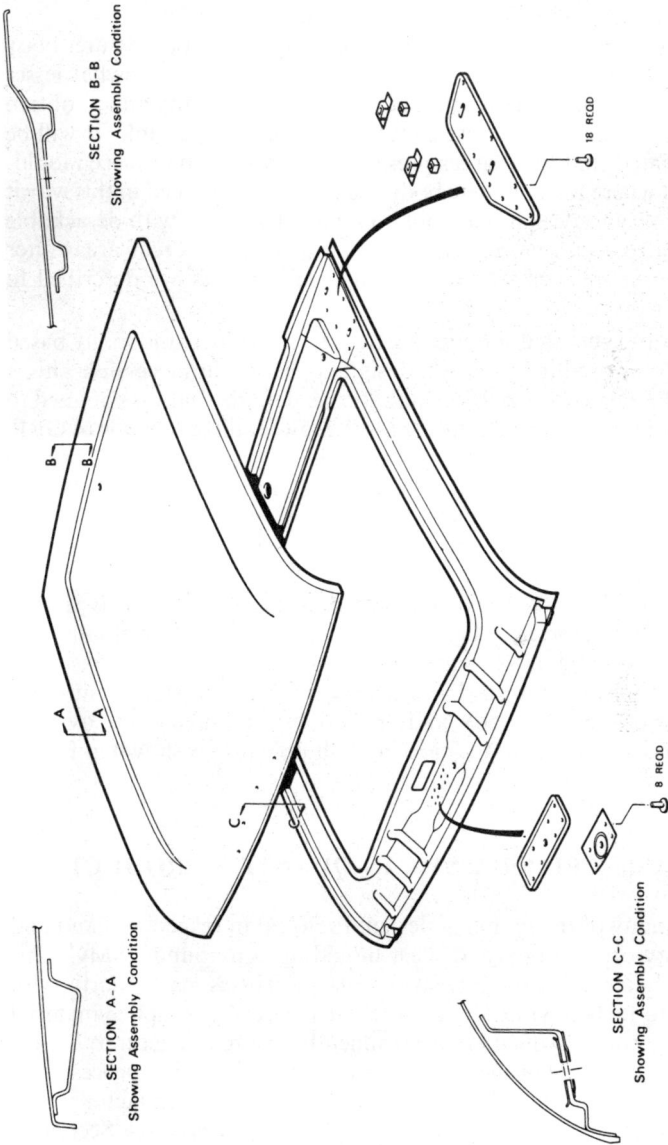

Fig. 5.18. The complete assembly of a double-skinned bonnet. (Courtesy Pressed Steel Fisher BL Components Ltd)

met with when using stiffening ribs and thus to avoid any danger of 'show-through', the inner skin should be applied as a secondary operation and after full cure of the outer component.

As in the pressed steel body the necessary return of the outer body skin at door, bonnet and boot openings, and to a somewhat lesser extent at screen and rearlight openings, confers rigidity and to obtain the maximum benefit these should be as deep as possible. It will be appreciated that these areas present problems in a one-piece moulding and where the complete body shell is to be produced in this way it is necessary to design the mould on multi-part lines with detachable sections that can be removed separately from undercut areas after cure. Some methods of constructing such moulds are described in Chapter 10.

Various types of door opening return can be used and many based on sections used in steel body design appear quite successful. This is due in large measure to the fact that considerable shape is confined to a small area thus compensating for the lower stiffness factor of GRP.

Early body designs incorporated the simplest of door opening returns and in consequence suffered from a number of problems such as distortion of the opening when under way, dropping of the door after a period of use due to inadequate rigidity at hinge points and shuffling and subsequent rubbing of both body and door. As techniques of mould design and moulding progressed body designers appreciating these early failures gave considerably more attention to dynamic loading at door areas and incorporated complex sections such as those used in steel bodies and ensured that stresses at hinge points both in the body and the door were taken out over an adequately large area. A typical door shut section is shown in Figure 5.19.

5.6   BASIC CRITERIA: SMC IN QUANTITY PRODUCTION

While many of the points of design discussed in respect of hand and spray lay-up also apply to sheet moulding compounds (SMC), this material has inherently different characteristics particularly with regard to its behaviour in the mould. First, it is a complete material the properties of which are determined by the manufacturer. Unlike the components of resin and reinforcement used in other processes its characteristics cannot be altered by the operator — a factor which contributes very largely to consistency in the end product. Secondly, SMC provides higher reinforcement-to-resin ratios than can be obtained by hand or spray lay-up processes and thus is most suitable for the production of components such as grilles and bumpers.

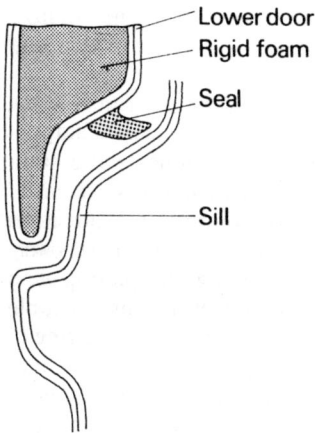

Fig. 5.19. *Typical section through sill and door shut*

Labels: Lower door, Rigid foam, Seal, Sill

Essentially it is a quantity production medium requiring the use of precision matched metal tooling and hydraulic presses and thus is confined in use to long production runs. It can be moulded in simple or complex shaped moulds and has the capability of reproducing fine detail.

The selection of SMC for a body component will follow similar lines to those applied to a GRP part produced by any of the other methods of processing with regard to the establishment of functional requirements. These will include structural and environmental factors, appearance, materials compatability and assembly requirements. Related factors to be considered will include the surface finish required, the capability of being self-coloured, chrome plated, machined and the ability to meet any safety requirements involved.

Economic viability will depend on the total contribution the material makes to the end use and not simply on a comparison of the material cost alone. Here questions of parts consolidation and simplification of assembly must be answered together with estimates of primary tooling and assembly fixture costs.

5.7 PROTOTYPE PROCEDURE

Having established the suitability of SMC for the component, the following operations complete the design process. Sketches of the part are made based on design flexibility, including wall thickness, draft angles, position of ribs, bosses and radii. At this stage there should be consultation with the mould maker to agree the feasibility of the design from a moulding point of view and economic factors including material, labour, primary tooling, assembly fixture and attachment costs must be determined.

Detailed drawings follow with the determination of critical stress points, wall thickness, ribs and attachment points. The next stage is the development of a prototype which should duplicate the component as closely as possible. Leading manufacturers of SMC materials recommend the use of partially completed production moulds for prototype development. Basically this involves 'over-building' the tooling so that prior to hardening the plating it is possible to modify the mould. Many body manufacturers use castable zinc based alloy tooling for prototype work. Moulding and testing of prototype mouldings with step by step refinements results in the final form of the component. The prototype components are tested under controlled conditions and the moulding variables of heat, pressure and cycle times are checked to determine the conditions for ensuring reproducibility on production runs. As mentioned the first production runs can be made on partially finished tools which allows reworking for changes in wall thickness and the adding or removal of ribs and bosses.

5.8 SPECIFIC DESIGN REQUIREMENTS

All surfaces, including ribs and bosses must be given a minimum of 1° relative to the opening plane of the mould. When undercuts are incorporated around the periphery of mouldings, by the use of corepulls and sliding inserts the movement of the corepulls and slides must be taken into account when determining draft.

Another 1° of draft is added for every 0.025 mm of texture depth when this type of finish is used in order to avoid damaging the surface as the part is ejected. The draft on right angle intersections on exterior surfaces can be determined by visually tilting the component to an increased angle and adding the necessary draft to the remaining sides to compensate for the moulding angle change on ribs and bosses.

Design thickness should normally range from 2.25–7.50 mm with corner radii of at least 1.5 mm to ensure material flow under moulding pressure and to minimise the occurrence of brittle, resin rich areas.

Due to the flow properties of SMC, strength can be obtained by the skilful incorporation of ribs rather than by merely increasing wall thickness. Ribs can serve as reinforcing for high stress areas, as mounting points or as partitioning. To achieve maximum strength a generous radius should be used where the rib joins the main panel. If possible ribs should be located behind styling lines, surface breaks or

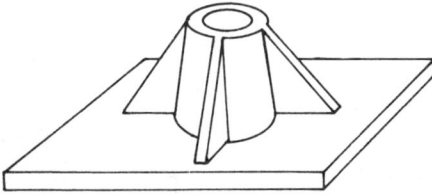

Fig. 5.20.   Preferred design for boss in SMC component

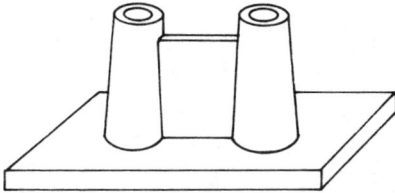

Fig. 5.21.   Linked bosses improve strength and aid material flow

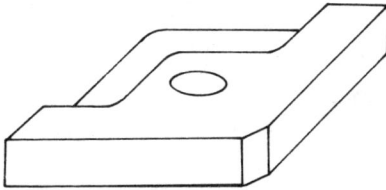

Fig. 5.22.   Corner bosses retain constant part and boss thickness

behind textured areas in order to mask the 'sink' in the surface of the component at the point where the rib and radius join the under surface. The use of 'low profile' SMC formulations (see Section 4.9) can reduce this tendency to depress the opposite surface which, of course, will also apply to bosses. The base width of ribs should not be less than 3.5 mm and not more than 10 mm. The base width of a particular rib will be determined by the width at the top which should not be less than 2 mm, with the requisite draft.

Holes perpendicular to the moulding angle can be formed with stationary plugs in the mould to reduce the need for post moulding operations. Undercuts and holes generally parallel to the moulding angle should be avoided where possible as their production increases mould complexity and cost through the need for movable cores. These devices can also be used to core bosses, so reducing material usage, and to form holes to accept mechanical fastenings.

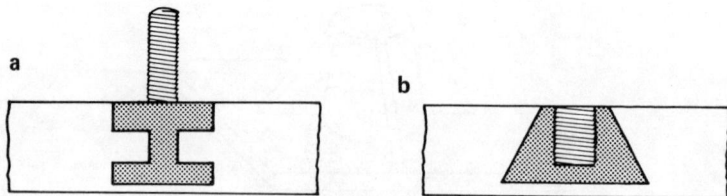

Fig. 5.23. Two types of moulded-in insert

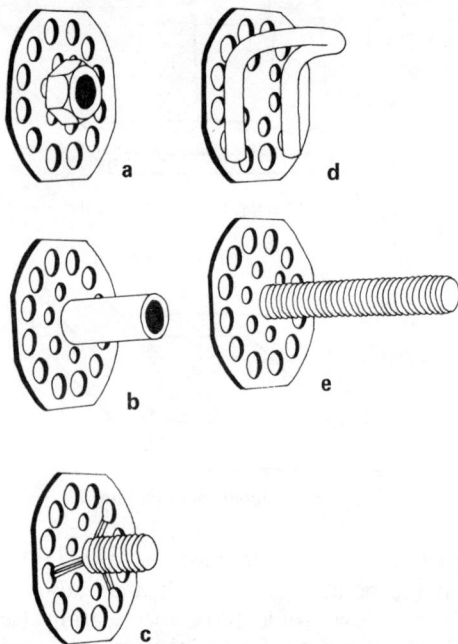

Fig. 5.24. The 'Bighead' insert system

The wall thickness of bosses should be about the same as the wall thickness of the component or not more than twice the diameter of the hole. Single bosses situated on a flat area should incorporate integral fins both for support and to maximise material flow during moulding (Figure 5.20). Where two or more bosses are adjacent to each other, they should be linked to improve strength and, again to improve material flow (Figure 5.21). Corner bosses, where possible, should also retain a constant part and boss wall thickness (Figure 5.22).

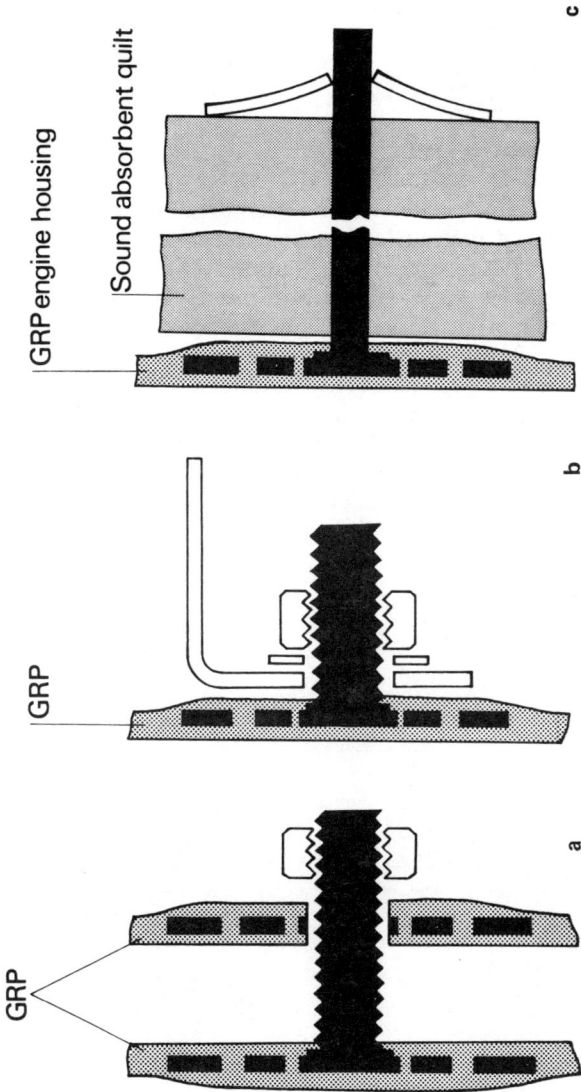

Fig. 5.25. Various methods of using 'Bighead' inserts

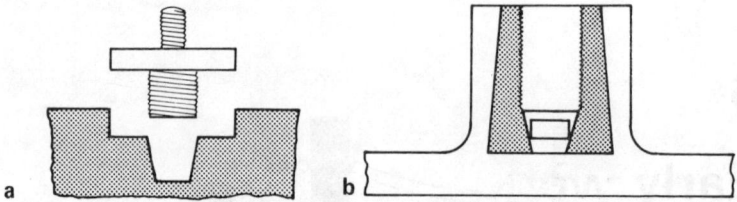

Fig. 5.26.   Two types of driven-in insert

Inserts play an important role in SMC parts. They can be moulded-in which takes up a certain loading time or can be pushed or driven-in as a post moulding operation. Figures 5.23(a) and (b) shows to commonly used types of moulded-in insert. A specially designed, highly versatile proprietary insert system known as 'Bighead' can be used (Figure 5.24 (a), (b), (c), (d) and (e)) in a number of ways. Some are shown in use in Figure 5.25 (a), (b) and (c).

Driven-in type inserts require high shear threads to achieve a suitable key (Figure 5.26(a)). The push-in type with knurled edges are locked in place by an inner brass slug which is driven to the bottom of the hole (Figure 5.26(a)). A second type is expanded to grip the bore of the hole by the insertion of the screw.

6

Early work

6.1 EARLY DEVELOPMENT WORK IN EUROPE

No doubt much of the early development work on reinforced plastics bodywork in Europe can be attributed to Auto Union[3] in Germany. Work commenced in 1938 when Auto Union, in collaboration with Dynamit AG, experimented with a crêpe paper reinforced phenolic resin as a body construction material without infringing patents covering the construction of steel bodies held at that time by Budd & Co. of the U.S.A. The mouldings were produced using 15,000 ton presses and chromium plated steel moulds. Material was loaded into the moulds in the form of bundles and tailored sheets. The roof was produced in one piece. The body was basically unstressed, being mounted on a rigid box-type chassis which was designed to absorb all the stresses when on the road.

The results of various tests to determine the behaviour of the bodies under impact conditions showed that they were very resilient and tended to bounce and crack rather than distort as in the case of a steel body shell. However, the nature of the paper/phenolic laminate was such that cracks were sharp edged and dangerous and the effect of the rebound in the event of an accident could cause considerably greater injury to the occupants than would be suffered in a comparable accident with a steel body. The project was dropped on the outbreak of war in 1939 but it had paved the way to the introduction of reinforced plastics body construction after hostilities. In 1949 experimental work was recommmenced at Auto Union with the building of a very simple body shape. A model, produced in soft wood, was covered all over with a 2.4 mm thick layer of Plasticine over which a hard plaster mould was cast. The Plasticine was removed, the plaster mould layed-up with glass rovings and resin and the model replaced forming a male plug. The resultant body had a reasonable finish on both inner and outer surfaces, a flexural strength of between 150 and 200 N/mm^2 and an impact strength of between 50 to 100 N/mm^2. Flammability tests, however, were not entirely satisfactory.

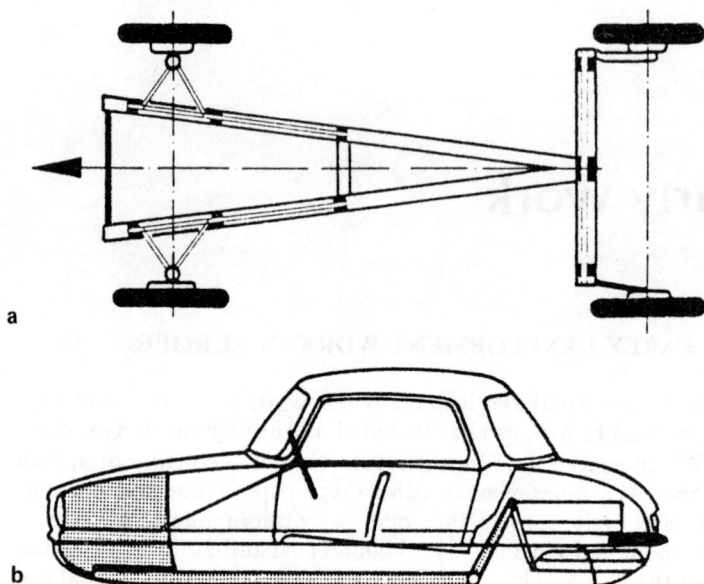

Fig. 6.1.   Early chassis by Auto Union for GRP body. (Courtesy Auto Union: GFK)

This work encouraged the company in 1953 to design a light 4-seater car which weighed 350 kg and which was successfully tested over a considerable mileage. The chassis was designed as a rigid structure with steel wheel arches so that the body was unstressed. The prototype is shown in Figure 6.1 (a) and (b). The final design, Figure 6.2, carried the use of steel reinforcement further, while still retaining the principle that all metal components were of simple form and that no pressings were to be used. Due to re-formation of the company in 1956 work was suspended but the experience built-up by the project was used to advantage in subsequent production of the company's steel bodied cars.

In the U.K. at this time, a number of amateurs were constructing glass/polyester bodies in simple moulds using the hand lay-up process. The majority followed conventional lines, and in the main were confined to small two-seat open sports type bodies intended for fitting to an existing chasses. Many failures were experienced largely due to a lack of appreciation of the need for controlled conditions, not only in the moulding shop but also in resin mixing and mould preparation and the need for an acceptably high glass to resin ratio. Another area in which failures occurred was due to the greater flexibility of the glass/resin laminate when compared with steel or

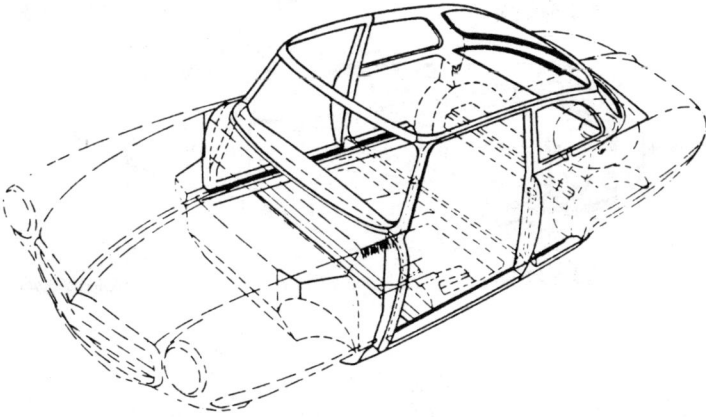

Fig. 6.2. *Later design by Auto Union with steel reinforcement.* (*Courtesy Auto Union: GFK*)

aluminium. This was highlighted by the fact that a number of these early bodies were produced from female moulds actually made off an existing body shell and thus did not incorporate sufficient double curvature to provide rigidity over the large, generally flat areas or sufficient stiffness and strength, at door points and screen mountings.

6.2   EXPENDABLE MOULD FOR PROTOTYPING

At this time the cold cure, hand lay-up technique was being applied, particularly in the U.S.A., for the production of one-off prototypes using an expendable plaster male mould. This method of prototype building permitted a manufacturer to produce a body shell for the purpose of obtaining corporate decisions on questions of styling, seating, colour, etc. both quickly and cheaply. Earlier, these questions were decided in the case of a new model by constructing a prototype in clay or plaster. Both these media, however, had the disadvantages of being very heavy, fragile and difficult to transport. Problems also arose in obtaining an accurate representation of colour and of plating on items such as bumpers, handles and screen surrounds. The glass reinforced polyesters have been found ideal for the building of prototypes as the body form presents an accurate visual representation at low cost and causes no problems in transport. Additionally, it can be flatted and sprayed in successive colours and items such as bumpers and window frames can be plated without difficulty to give an authentic appearance.

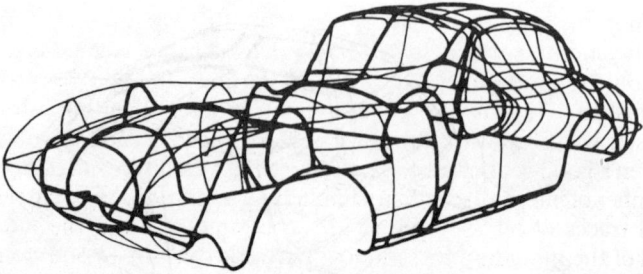

Fig. 6.3.   Loft lines reproduced in steel tube for retention in the finished, male moulded body

The method was not confined to the construction of design studies but was also used for one-off bodies such as produced by the Author in the early 1950s. A brief description of the method is given in the following section. In the light of experience a number of modifications would now be made but the general process used still holds good for a single body.

6.3   ONE-OFF MALE MOULDED COUPÉ BODY

The close-coupled coupé body was designed with flowing lines, much appreciated during the period, and with as much double curvature contouring as possible. Because of the method used for moulding the complex shapes there was no need to make any concession to the problem of removing the finished body from the mould, as would have been the case had a one-piece female moulding technique been applied.

The first stage was the building of an accurate scale model in plasticine from which a multi-part female mould was cast in plaster of Paris. From this mould another model was cast, also in plaster, from which the loft lines of the body were taken at fixed stations along the model.

For the construction of the full scale mould, these lines were reproduced in drawn steel tubing which in this case remained in the finished shell to act as stiffening members (Figure 6.3). To facilitate plastering the expendable mould, thin rods which conformed to the body lines in both transverse and longitudinal direction were lightly welded to the main tubes and were covered on the inner side with fine mesh wire netting. During plastering attention was given to the thickness of the mould as this needed only to be thick enough to withstand trimming to final form and sanding to produce a reasonably smooth surface. It will be appreciated that when using a male mould the surface is on the inside of the body and does not have to be finished to the same degree as a female mould. When smoothed, the plaster was filled with successive coats of Shellac which when dry was given a good coat of wax followed by a PVA release agent. Due to the number of internal, secondary laminating operations to be carried out, all traces of release agent had to be removed from the inner surface of the moulding by sanding in order to ensure a good bond with internal additions to the moulding. In this way, when the laying up of the body shell was completed and the resin fully cured, the unwanted plaster, thin rods and the wire netting were knocked away from the inside leaving the shell in contact with the main stiffening ribs. The inner surface of the laminate was cleaned and sanded in the area of the tubes which were then over laminated with resin impregnated glass cloth to embed them completely in the body shell.

In this project, bonnet and boot lid were cut from the single piece body moulding. Stiffening was achieved by the use of balsa wood profiles over-laminated with impregnated cloth on the underside. The doors were completed and stiffened by the bonding of a shaped inner panel which had the effect of producing a box section with space between the inner and outer panels to house the window winding mechanism.

Contact and weathering faces for both bonnet and boot panel were produced by replacing the cut and stiffened panels in the body and taping their outer surfaces flush with the surrounding body surface. The under surface around the periphery of the panel was coated with release agent and strips of impregnated glass cloth were laid over the junction of the panel and body shell. When cured the panel was removed and the permanent under fillet trimmed around its inner edge. Thus, when the panel was replaced its outer surface was flush and followed the contour of the body moulding more accurately than most current production bonnet and boot lids.

The chassis selected was by Lancia and was extremely rigid having been designed to carry an original open body which, built in the conventional manner of ash framing and steel panels, was unstressed. The GRP shell was mounted at six points, three each side, by means of Silentbloc rubber bushes and through bolts from the large diameter tubes which formed the lower edge of the body shell. This system not only gave a very slight resilience at the mounting point to reduce the transmission of noise but also permitted rapid removal of the complete body to facilitate service. The only other attachment between body shell and chassis was through two face plates welded to the scuttle and body respectively. A feature of the design necessitated the instrument panel, steering column support, etc. to be mounted on the chassis. In this way the body could be removed leaving a drivable chassis. All electrical connections from chassis to body were provided with plug-in connectors. All windows and screen surrounds were of aluminium channel-section moulded into the body shell.

Finishing of the shell was carried out by sanding and filling with a polyester putty after which the body was primed and cellulosed in the normal manner.

6.4   EARLY PRODUCTION BODIES

At about this time a number of companies had commenced limited production of GRP bodies with some success. Designs were not confined to open bodywork and lessons were being learned regarding the incorporation of double curvature to impart rigidity and the use of moulded-in steel or aluminium reinforcement at stress points such as body mounting points, screen mounting areas and, most impor-tantly, at door hinge and lock points. Among pioneers a number of which are no longer building bodies were Rochdale, TVR, Turner, Reliant, Bond, Berkeley, Lotus and Shamrock, to name but a few. The Rochdale GT bodies, built in Rochdale, Lancashire, in the early 1950s were moulded, as were all bodies at that time, using hand lay-up methods. They were sold largely as a body for amateur fitting, to almost any chassis, with modifications, and were available as a Type 'F' or Type 'C'. Later the Company produced the Type 'ST' designed for fitting direct to the 8 or 10 Ford Popular.

In 1956 a 2 plus 2 saloon body was produced which was supplied complete with wheel arches, bulkhead and with windows and lockable doors fitted. The one-piece body shell was found to add considerable rigidity to the widely used Ford chassis frame. This success led to the design of a new body with a GRP floor and with tubular steel stiffening at critical points using Morris Minor suspen-sion units. Experience with this body and the success of the monocoque Elite then being moulded for Lotus by Bristol Aircraft, encouraged the design of the Olympic which went into production in 1960. It was a complete self-coloured body shell suitable for fitting by an amateur constructor with Morris Minor, Riley 1.5 litre or Ford 8 or 10 components. It featured a one-piece moulding with undershield, and bulkheads, dashboard, wheel arches and rear shelf bonded-in while the main body shell was still in the mould.

Following a serious fire in 1961, the Racing Car Show in London in 1962 saw the introduction of the Olympic Phase 2 with Ford Classic 1.5 litre or Cortina engines, disc brakes and Triumph front suspension. The new body had an opening rear window and accommodation for twin fuel tanks which allowed the spare wheel to be fitted beneath rather than above the rear platform giving increased luggage space. The Phase 2 when fitted with the Cortina GT engine giving 78 bhp had a maximum speed well in excess of 110 mph. Some 250 of the bodies were built before production virtually ceased in 1973.

Another example of a new car project based on the economics of using a hand laid-up GRP bodywork which was symptomatic of the period was the American backed Shamrock. Unfortunately this project, unlike the Reliant, TVR and Lotus, did not survive although a limited number of cars were completed in Ireland, the venue originally intended as the production base. The concept of the Shamrock was an open four-seater of modest performance, but with adequate passenger and luggage space and styled very much on contemporary American lines. Production of the full-scale master model and piece-part female mould in GRP was undertaken by the author using two highly specialised plasterers accustomed to working in plaster of Paris. The building of master and female mould followed the lines described in Section 10.3 and when complete was mounted on a special chassis designed by consultant L. Ballamy. The mould and the first body were produced in polyester resin supplied by British Resin Products and following the car's first public appearance all moulds and production tooling were eventually transported to Ireland where a limited run of cars were produced by S. Rhiando.

The TVR sports coupé, one of the very few early projects which became established and is produced in highly sophisticated form today to be numbered among the U.K.'s high performance cars, commenced production in 1947. The first cars, built in Blackpool works, were fitted with alloy bodies which shortly after were replaced by GRP body shells moulded by Rochdale Motor Panels Ltd. In 1957 the current model was shown in New York and models were subsequently exported for sale by the U.S. distributors R. Saidel. The success of the design led to the production of the company's own GRP body on the 'Grantura' Mk II and later 'Griffith' 200 and 400. Production in 1964 was of the order of 10 cars per week and at the end of 1964 something approaching 200 'Griffith' V8's had been built. The majority were exported to the U.S.A. where some were fitted with modified bodies by their owners. Notable was the fact that this level of production was obtained with a labour force of 100 at the Grantura plant.

Later TVR models followed. Bodies were moulded in two sections, a floor pan and an upper structure both of which were bonded together with a number of minor mouldings. Once removed from the

Fig. 6.4.   Saloon car body in GRP. Door outer panel mouldings. (Courtesy Pressed Steel
Fisher BL Components Ltd)

Fig. 6.5.   Saloon car body in GRP. Door inner panel mouldings. (Courtesy Pressed Steel
Fisher BL Components Ltd)

Fig. 6.6. *Saloon car body in GRP — moulded body components. (Courtesy Pressed Steel Fisher BL Components Ltd)*

Fig. 6.7. *Saloon car body in GRP — main body shell. (Courtesy Pressed Steel Fisher BL Components Ltd)*

Fig. 6.8.   Experimental car body in GRP. (Courtesy Pressed Steel Fisher BL Components Ltd)

mould and still minus the bonnet, the main shell was trimmed and fitted to a slave chassis/trolley. After fitting doors and with a separately moulded bonnet the assembly was passed through an oven two or three times at 60°C, and air bubbles were removed and filled. The bodies were given an etch primer coat to fill minute pores followed by six coats of primer, three coats of cellulose and checked for quality. After mating to its drivable chassis, windows, heater, wipers and servo were installed. In the trim area the body received sound-deadening trim, dashboard and seats. Lastly, the bonnet was fitted and the car road tested. At this time some fifty cars could be in course of assembly at any one time. Later sophistications included zinc plating of all suspension wishbones and an acrylic body finish.

Production figures were: 1972: 380 cars; 1973: 390; 1974: 420; 1975: 500. Today the success of the TVR, which commenced with such modest production, continues with the latest TVR 'Turbo' a 150 m.p.h. sports coupé in the best GT tradition. Illustrations of current TVR production models are given in Figures 7.6–7.11.

6.5   EXPERIMENTAL SALOON

Early work in the 1950s by Pressed Steel Fisher B.L. Components Ltd., included the development of a small experimental saloon body incorporating GRP doors and bonnet. The doors were designed as a

Fig. 6.9. Front view of experimental saloon body in GRP. (Courtesy Pressed Steel Fisher BL Components Ltd)

two skin construction. Figures 6.4 and 6.5 show the inner and outer moulding prior to assembly. The complete set of mouldings making up the body is shown in Figure 6.6. with a view of the dash panel assembled in Figure 6.7. Two views of the complete experimental car are shown in Figures 6.8 and 6.9. Details of the engine compartment and bonnet hinging arrangements are seen in Figure 6.10.

6.6 CHEVROLET CORVETTE

The Corvette built by the Chevrolet Division of General Motors from 1953 is unique as the first GRP bodied car to be produced in quantity. Having passed through many styling exercises the Corvette still holds pride of place being one of America's leading high performance sports cars, and in production today. Moulding of the early bodies was by a matched mould system using metal moulds. The complete body consisted of some 24 major mouldings which, when assembled, weighed 153 kg. This weight was accounted for by 61 kg of refinforcement, 69 kg of resin and 23 kg of inert filler. Tests indicated a tensile strength of 1050 kg/cm^2, from which it was calculated that a laminate 1.5 mm thick should be equivalent to a pressed steel panel

Fig. 6.10.   The engine compartment of an experimental saloon body in GRP. (Courtesy Pressed Steel Fisher BL Components Ltd)

1.0 mm thick. Other tests carried out to evaluate mouldings made of resins of different flexibility included the dropping of a 38 mm diameter steel ball from a height of 304 mm on the reverse side of the mouldings at ambient temperature. As a final check the mouldings were required to withstand a drop of 254 mm after they had been exposed to a temperature of 100°C for a period of seven days. Another critical factor in the specification was that the water absorption figure should not exceed $\frac{1}{2}\%$ after 24 hours immersion in water at room temperature.

General Motors reportedly estimated that the cost of the GRP body tooling was, at that time, approximately $500,000 as compared with about $4,500,000 for similar tooling for a steel body. In addition there was the advantage that in using GRP the body could be put into production very much earlier than could a similar body in steel.

7
Current work

7.1 INTRODUCTION

Since the early 1950s when specialist builders were welcoming glass reinforced plastics as a medium for producing bodywork incorporating the complex double curvature and flowing lines which were in vogue at that time, without the expertise and expense associated with panel beating, there have been considerable changes in the approach of the material manufacturer, body designer and public with regard to the application of the medium. Today GRP bodywork is no longer regarded as an inexpensive alternative to steel or aluminium. It is in many instances a preferred body material — a situation borne out by the high used prices of the few marques of GRP bodies cars produced by the *skilled* manufacturers on both sides of the Atlantic.

In 1978, the Corvette's twentyfifth anniversary year, Chevrolet produced some 50,000 Corvettes, a staggering figure compared with European output of GRP bodies passenger cars. In Europe the automobile industry, with the exception of its few giants, historically has been made up of a comparatively large number of small companies each of which, in the main, have produced an individual design with appeal to various sections of the market. This situation still obtains, albeit with fewer manufacturers in the field. Brief details of production by some companies representing different aspects are given in the following pages.

Among European manufacturers who currently build GRP bodies on a production scale are Lotus Cars who produce the Elite, Esprit and Eclat high performance sports saloons, Figures 7.1–7.5; TVR Engineering who, as described in Chapter 6, now produce one of the U.K.'s grand turismo cars in the best of the tradition, Figures 7.6–7.11; Panther Westwinds, designers and producers of a small range of highly successful two-seat open sports cars of nostalgic appeal and Reliant who in addition to its popular three-wheeled Robin, are the manufacturers of the Scimitar sports saloon. In France, the highly successful Matra-Simca 'Bagheera' has a dedicated following.

Fig. 7.1.   The Lotus Elite, 4-seat. (Courtesy Lotus Cars Ltd)

Fig. 7.2.   The Lotus Esprit, 2-seat, mid-engine. (Courtesy Lotus Cars Ltd)

7.2   GRP MINI AND MG1300

The application of hand lay-up GRP, whilst having disadvantages from a quantity production point of view due to its inherent labour intensive characteristics has for this very reason a considerable appeal to a manufacturer in a country where labour is plentiful, inexpensive but in the main semi-skilled. This advantage has been exhibited by Chile where since 1968 a local manufacturer has been producing a GRP bodied version of the Austin Mini at rates of approximately 100 per week. The body, Figure 7.12, was developed by Pressed Steel Fisher BL Components Ltd., using the hand lay-up technique with

Fig. 7.3. An Esprit body shell removed from the mould and progressed to fettling for 'fine trimming'. (Courtesy Lotus Cars Ltd)

Fig. 7.4 The Lotus Esprit production line (note protective GRP covers on the wings). (Courtesy Lotus Cars Ltd)

Fig. 7.5.   The Lotus Eclat, 2+2 seat. (Courtesy Lotus Cars Ltd)

Fig. 7.6.   TVR production. Hand lay-up of body floor pan prior to mould sections being bolted together. (Courtesy TVR Engineering Ltd)

Fig. 7.7.   TVR Taimar body shell removed from the mould and being fettled to correct dimensions from mould scribe lines. (Courtesy TVR Engineering Ltd)

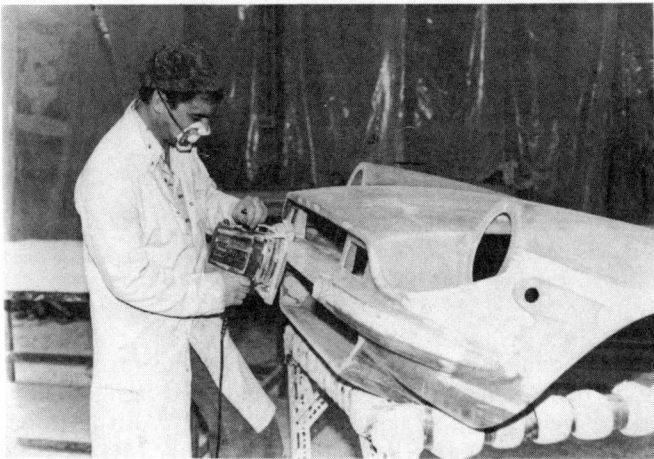

Fig. 7.8.   TVR bonnet being prepared for priming. (Courtesy TVR Engineering Ltd)

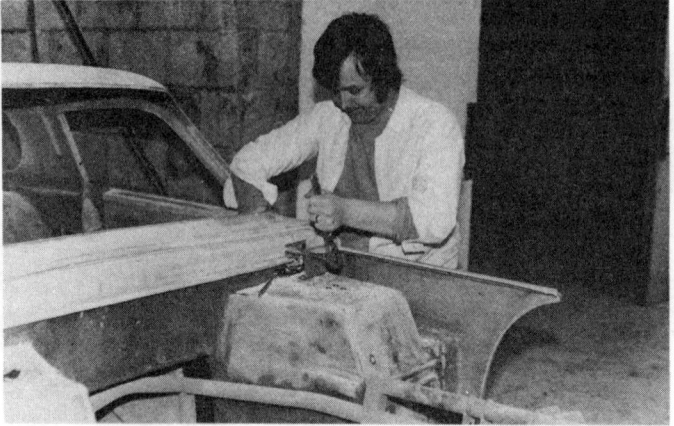

Fig. 7.9. *Fixing the bonnet catch mechanism on a TVR body. (Courtesy TVR Engineering Ltd)*

Fig. 7.10. *Installation of painted body and 3 litre, V6 power unit in a stove enamelled steel TVR chassis. (Courtesy TVR Engineering Ltd)*

the body mounted on a sub-frame at front and rear and with a birdcage body stiffening structure. Moulding is facilitated by the absence of a gutter and 'straight' side on the body. Also developed by Pressed Steel Fisher is a GRP version of the popular MG 1300 saloon, Figure 7.13, for production in developing countries.

Fig. 7.11. The TVR range comprises the 3000M Taimar, shown here, and the convertible. (Courtesy TVR Engineering Ltd)

Fig. 7.12. Special version of Austin Mini incorporating GRP bodywork for manufacture in overseas developing countries. (Courtesy Pressed Steel Fisher BL Components Ltd)

7.3 LOTUS CARS

For many years Lotus used hand lay-up methods, but with increased production and the need for greater consistency now use their own development of the resin injection system. Reportedly, some few years ago the company installed a Resinject unit capable of injecting resin at a rate of approximately 5 kg/min which as far as could be ascertained was used to produce all the detachable panels fitted to the

Fig. 7.13.   Special version of the MG1300 incorporating GRP bodywork for manufacture in overseas developing countries. (Courtesy Pressed Steel Fisher BL Components Ltd)

then current Europa. The fact that the unit was not used to mould the body shell itself was understandable and in all probability due to the fact that it is not usually practicable to modify a one-piece moulding intended for hand lay-up to a matched mould system.

When Lotus introduced the four-seater Elite, to be followed by the Eclat and Esprit range, export sales demanded that the designs meet U.S. safety regulations. Thus, to maintain the car's performance it was necessary to find a way of accurately controlling the glass/resin ratio and wall thickness of the mouldings. The use of the resin injection system would have met these requirements but at considerable mould cost, and possible acceptance of a weight penalty. It is in the area of resin injection that the company have been so successful in developing the system. By designing their bodies as two-piece mouldings with the joint horizontal along the waistline it was not only possible to simplify each moulding but also by the use of only three moulds, one lower and two upper, to produce two body styles.

Lotus Cars produce some 25 Espirt and 5 Elite/Eclat models a week using a staff of approximately 60 in the moulding shop. Five sets of open moulds are used for Esprit production; one resin injection mould for the Elite/Eclat. The Esprit moulds are composite epoxide/-polyester each being some $5 \times 2.2 \times 1$ m in size and have an extremely high finish on the moulding surfaces.

7.3.1   Transfer painting

The system developed by Lotus for producing a paint finish is unique: the paint is sprayed into the mould after the application of the parting agent and *before* spraying of the gel coat. The paint, a polyurethane formulation used at a thickness of 0.1 mm, is cured at 37°C for three hours after which the gel coat is sprayed in. The system, known as 'transfer painting' is only used at present in conjunction with open moulds. Although the gand lay-up is in itself slower the system offers the advantage of 25 man hours saving on the full moulding cycle as it eliminates matting down the finished body to give a key for a conventional paint process: the finish is excellent. When the gel coat has cured, hand lay-up is carried out and the two main body mouldings are bonded whilst still in their moulds.

7.3.2   Resin injection moulding

The moulding of Elite/Eclat bodies in halves is now by means of a vacuum assisted resin injection process in which atmospheric pressure is used for mould clamping. Additionally, this ensures location of the mat against danger of drift during the injection cycle. After locating the mat, the mould is closed, vacuum applied and a specially formulated resin injected. The injection cycle reportedly takes about 20 min and cure just over an hour. Originally the company intended to use hollow sections in the Elite body in place of the foam now incorporated. There are a number of ways of doing this, notably by the use of an inflatable member. However, the problems of ensuring against puncture of the inflation member and the difficulties associated with joining the hollow mouldings, and of quality control, has led the company to retain the foam coring. There is very little cost or weight penalty incurred by the use of foam and the additional rigidity afforded by the construction is considerable.

7.4   MATRA-SIMCA 'BAGHEERA'

The now well established Matra-Simca 'Bagheera' sports car built in France is another example of the few GRP bodies cars produced by a resin injection process. Until the 'Bagheera' was designed companies wishing to use a resin injection system had difficulty in finding equipment capable of handling filled resin systems, catalyst pastes at high percentages, high back pressures and the pressure variations due to the shape of different moulds on the production line. Additionally, there were problems of injecting precise and pre-selected volumes into each mould which necessarily were often some distance from the equipment. However, prototype equipment specially built by the U.K. company, Liquid Control, proved highly successful and a

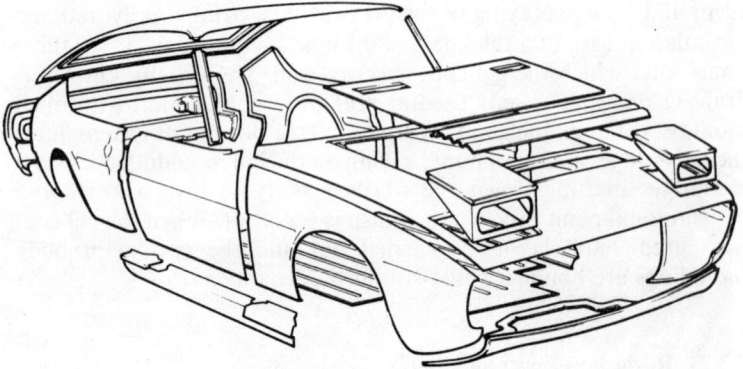

Fig. 7.14.   Exploded view of the Matra-Simca 'Bagheera' showing the components of the body produced by resin injection

number of their machines are currently being used at Matra-Simca.

The 'Bagheera' is an excellent production example of the application of high strength, low weight body mouldings with a high degree of surface finish produced by injection. As shown in Figure 7.14, there are 16 body components produced from a number of production lines. Each component is laid out in a circular manner to facilitate movement of trolley mounted moulds as they pass through the sequence of preparation, injection, curing, demoulding and painting. Early moulds were of epoxide resin but latterly high strength, wear resistant metal moulds with epoxide reinforcement are used. Clamping of the moulds is carried out in hydraulic presses.

7.5   PANTHER WESTWINDS PRODUCTION

A comparatively new company, Panther Westwinds Ltd., commenced production at Byfleet in the early 1970s and now produce some 10–15 Panther Lima and Turbo Lima GRP bodied cars per week. Initially the company, set up by Bob Jankel, designed and built the Panther J72 a highly successful 2-seater sports car in a style reminiscent of the SS100. Powered by a 4.2 Jaguar engine the car was fitted with an aluminium body built to the highest standards.

In August 1974, J72 production was joined by the Panther Ferrari FF. This car, seven of which were built, was based on the 330 Ferrari GTC chassis with its 350 b.h.p. V12 4-litre engine with the addition of Girling 'Formula One' coil spring damper units with adjustable spring rates and a tubular steel frame to provide a basis for the hand

Fig. 7.15.   The Panther Lima 2-seat sports car

built aluminium body. No doubt the culmination of design expertise and the high degree of skill of their body engineers is in the Panther De Ville, a luxury saloon or convertible in the style of the great cars of the 1930s.

Panther Westwinds entry into the GRP body field was basically for economic reasons. Expansion demanded production of a relatively inexpensive, attractive and readily serviced model. The current Lima, Figures 7.15 and 7.16, is the result of a full appreciation of the advantages and shortcomings of GRP as a body material combined with a very shrewd choice of standard and readily available mechanical components. Avoiding the pitfalls associated with a stressed structure the Lima's body, Figure 7.17, is completely unstressed clothing an extremely rigid box-chassis based, in the earlier models, on a Vauxhall Magnum underpan component and later, in the Mk 2 models, redesigned and built to the company's order. Engine and running gear are Vauxhall Magnum, with the standard power unit giving 108 b.h.p. A turbo charged version is available giving some 180 b.h.p.

All bodies are to the same design, the only variation on models destined for the United States being left-hand drive with a bumper modification to comply with U.S. regulation in this area. The body is produced by hand lay-up methods and is moulded in two sections, fore and aft. Mounting of the shell on the chassis is by means of self-tapping screws at the rear, along the sills and by bolts at two points

Fig. 7.16.   The Pantha Lima fitted with optional hard top and luggage rack

Fig. 7.17.   The body moulding of the Pantha Lima

beneath the front wings, thus permitting the body to be removed leaving a drivable chassis. With an appreciation of the problems raised by GRP doors, the Lima uses standard MG steel units complete with wind-up windows. The doors are hung from the rigid scuttle structure thus avoiding any tendency to distortion, shuffling or dropping in service.

Fig. 7.18. The Pantha Lima assembly line

Fig. 7.19. The 'Equus' specially built by Panther Westwinds for Vauxhall Motors

After mounting on the chassis the bodies undergo a baking process followed by a very thorough check for any moulding faults. The bodies are rubbed down, primed and a very complete cellulosing process is followed involving some 20 different coats. Assembly of mechanical components is carried out in an airy and spacious shop, Figure 7.18.

In addition to production of the J78, De Ville and Lima cars, skills at Panther Westwinds are much in demand by the large manufacturers for the construction of special and development projects. One of the latest to be completed by the company is the Equus, Figure 7.19, specially built in GRP by Panther for exhibition by Vauxhall at the 1978 Motor Show.

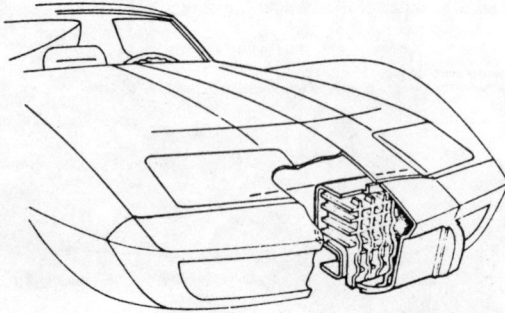

Fig. 7.20.   Chevrolet Corvette front end treatment

7.6   CHEVROLET CORVETTE PRODUCTION

Development of the Corvette up until 1970, covered briefly in Section 6.5, was confined to wet lay-up mouldings and it was not until after this date that a start was made in converting production to SMC moulding.

Prior to 1970 the company had been evaluating the development of SMC but fully appreciated that very often there was no advantage to be gained in changing from a component designed for wet lay-up to SMC. The move could result in a cost penalty. Some Corvette components, however, did lend themselves to SMC and a compromise was made by producing the rear end of the body in the new material. Experimental work pointed to the need to re-tool in order to meet specification and it was not until 1973 that the Corvette outer panels were 100% low profile SMC.

The first components to be produced in low profile SMC were the plenum centre panel, exterior rear quarter panels and the rear end panel. Detailed studies pointed to the need for important changes in SMC production. These included improvement in the buck-bond fixing to relieve as much built-in joint strain as possible, to sand blast rather than sand the bond joint area and to revise the resin system to obtain improved bond strength. This was done and the car was produced to the complete satisfaction of the engineering and plant personnel.

In the 1975 model, as in current Corvettes, the headlamps are automatically retracted when not in use. Figure 7.20 shows the special front bumper system introduced to comply with the 5 m.p.h. impact legislation and with the headlamps in the retracted position. Figure 7.21 shows the later type front end with a cellular cushion of one-piece impact modified polyethylene selected to ensure good impact absor-

FRONT BUMPER COMPONENTS

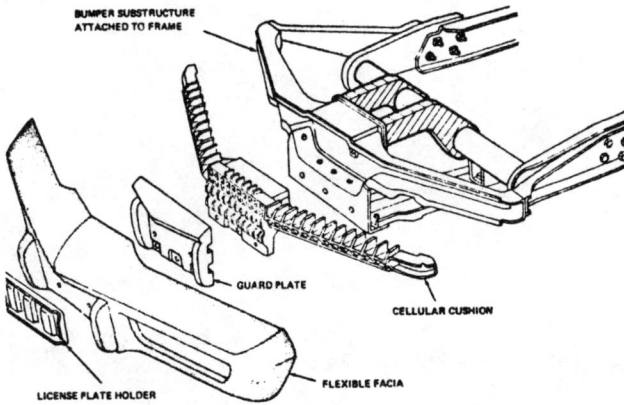

Fig. 7.21.   Later type front end on Chevrolet Corvette

REAR BUMPER COMPONENTS

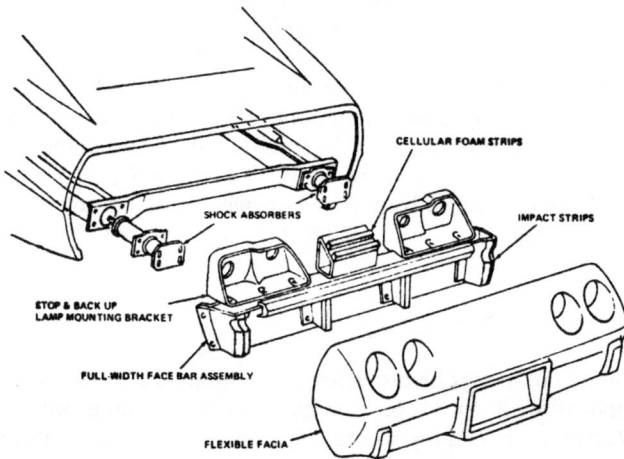

Fig. 7.22.   Later type rear end on Chevrolet Corvette

ption under repeated impacts of up to 5 m.p.h. The rear bumper arrangement designed to fulfil a similar function and using Delco hydraulic-pneumatic energy absorbers is shown in Figure 7.22. This necessarily brief description, which outlines some of the special requirements of Corvette production, gives no indication of the thousands of man-hours invested by the company in the application

Fig. 7.23.   Large front end mouldings in SMC. (Courtesy Owens-Corning Fiberglas Corporation)

of GRP to this unique car. Similarly for reasons of space it cannot detail the exhaustive tests carried out by the company in its search for an optimum bonding and painting system, irrespective of the supplier of the resin used. To quote R. A. Vogelei of the Chevrolet Motor Division, 'Taken collectively these stringent compatibility requirements pose the single largest problem facing us to-day'.

7.6.1   Weight reduction

An example of the problems facing U.S. car manufacturers regarding compliance with the U.S. Energy Act are brought into sharp perspective as they apply to the Corvette. The 1978 Corvette has a curb weight of approximately 1589 kg. High usage options can add another 68 kg so that with a 136 kg passenger load the car is in the 1816 kg inertia weight class. Chevrolet's goal by 1982 is to reduce the weight of the Corvette by 13% or 204 kg to bring the car into the 1589 kg inertia weight class. At a constant power weight ratio it is calculated that this will lead to an improved fuel economy of 2 m.p.g. GRP materials are expected to contribute 68 kg to this total with about 60% coming from existing GRP parts and the other 40% from new applications of GRP.

Fig. 7.24.   Front end on Chrysler designed in SMC. (Courtesy Owens-Corning Fiberglas Corporation)

7.6.2   Finish

Another of Chevrolet's highest priorities is improvement in the exterior finish of the SMC components. A major step in this direction will be taken with the application of in-mould coating to all exterior body panels. The process has been developed jointly between General Motor Mfg Development and General Tire. Tooling modifications to incorporate in-mould coating are being applied first to roof panels. For the 1980 model all exterior panels will be so treated. It is expected that in-mould coating will confer a number of advantages, ie. the elimination of porosity and surface defects, greatly reduce sinks opposite bosses and ribs, simplify paint priming operations and allow the use of glass bubbles as a filler material. The use of the latter material as a filler in place of conventional talc, if successful, would save an additional 49 kg in weight.

Other applications being evaluated for the early 1980s models with anticipated weight savings are (1) a removable roof panel with integral ribs and attachment bosses thus eliminating the conventional inner reinforcement, $4\frac{1}{2}$ kg, (2) a GRP spare wheel developed from the concept wheel of General Motors Mfg Development, $4\frac{1}{2}$ kg, (3) a GRP radiator support, 7 kg, and (4) some XMC material in bumper structural parts and a graphite and GRP rear leaf-spring, expected to effect a further weight saving of $13\frac{1}{2}$ kg.

Attention to weight saving is being practised in varying degrees by most of the U.S. manufacturers and, as discussed earlier in this section, not only to sports cars. Figures 7.23 and 7.24 illustrate the use

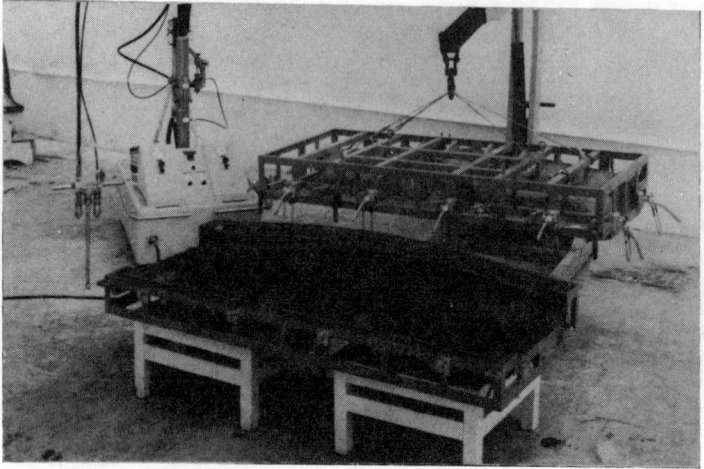

Figs. 25–30 show the sequence of operations in the resin injection moulding of a large vehicle panel. (Photographs by courtesy of Savid of Como and Coudenhove)

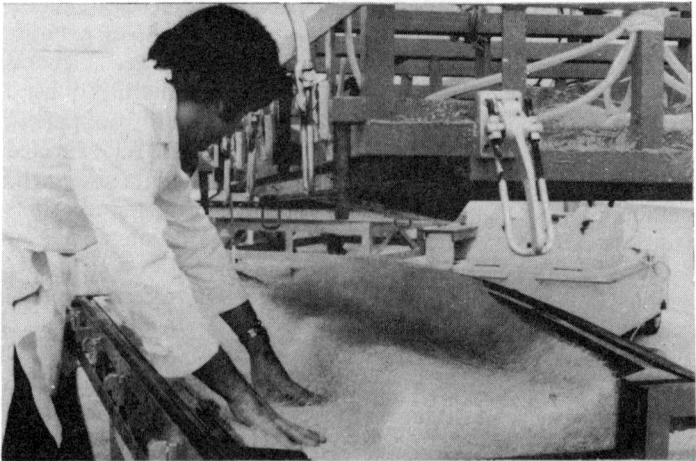

Fig. 7.26.

of SMC mouldings by Chrysler to name but one company. In Europe also the application of GRP is well under way for both small and large body components in other than purely passenger vehicles.

Fig. 7.27.

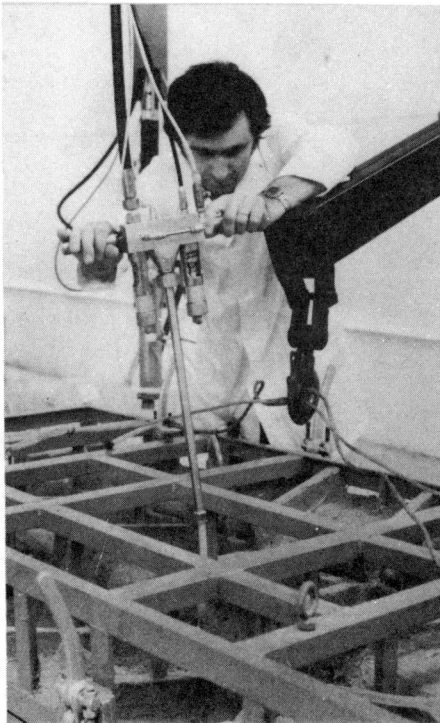

Fig. 7.28.

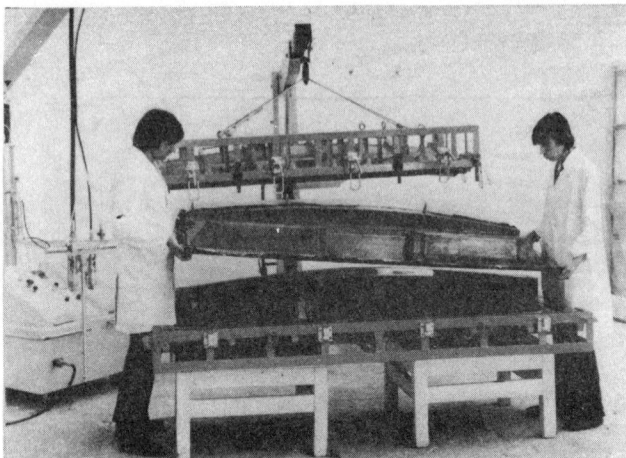

Fig. 7.29.

Fig. 7.30.

7.7   SAVID: RESIN INJECTION WORK

Considerable work is being carried out in Italy in the production of
large vehicle panels using the Coudenhove resin injection system and
moulds made by the Vienna-based company. Figures 7.25–7.30 show
the sequence of operations in the resin injection of a large automobile
panel which is now being produced to close tolerances both in weight
and dimension.

8

Future trends: quantity production

8.1 DESIGN OBJECTIVES

Having mentioned briefly some of the more basic points of design in hand spray-up and SMC moulding in Chapter 5, one can look at current and future trends in the large scale application of GRP materials to automobile bodies and body components. While some years ago many designers believed that the role of plastics, not necessarily GRP, was confined to a few secondary structures and minor components, largely in the electrical and ancillary areas, some had more foresight and as time has shown have been proved right. In 1957 there was only some 20 lb of plastics materials in the average U.S. passenger car, today there are between 150 and 200 lb, including GRP, in a comparable model and this quantity is expected to rise to between 350 and 500 lb by 1985. What then is the future trend for GRP? One important long term economic consideration which obviously does not concern the specialist builder but which is a factor that weighs heavily with a major manufacturer is the relative energy consumption of plastics and metals. The net energy cost of plastics is lower than steel and some four times lower than that of aluminium.

8.1.1 Weight reduction

Development work involves large exterior panels, roofs, boot and engine lids and includes the use of reinforcements other than glass fibres. Glass reinforced polyesters — particularly SMC — are at present making significant inroads into areas which hitherto have been viewed as the precincts of steel. Current examples in quantity production — an area where the considerations are very different to those affecting the amateur or small production body builder — include bumpers and front ends. While it is not within the scope of this book to discuss components other than bodies and associated components, it should be remarked that designers are even now carrying out experimental work on the use of carbon, graphite, and aramid fibres as reinforcement for structural members such as radiator supports, leaf springs, drive shafts and even road wheels. The important factor is to reduce weight and this is also one of the main aims of the body designer. As an indication of how far American designers have gone already, the average curb weight of a car in the U.S. today is 4300 lb. In 1985 it is confidently expected that this figure will be reduced to something in the order of 2600 lb. This reduction, aimed at a lower fuel consumption will undoubtedly be due to a far greater use of plastics in both engineering and body components.

Many of these forecasts as they apply to the body are discounted by some at this time on the grounds of material cost and labour content. Others believe that the all-plastics passenger body could be designed and mass produced successfully. Reportedly[4] plans could include a range of three models, a four-door saloon, a four-door station wagon and a two-door sports coupe. Careful design involving interchangeable panels would permit longer runs of common components and thus benefit production economies of the total range.

The market for a range of such cars is difficult to forecast but in view of the excellent condition of, for example, Corvette and Studebaker Avanti bodies after 14 and 23 years, respectively and the obvious appeal of a GRP body as reflected in the high prices being obtained for these models, it could be substantial. With the elimination of the rust bugbear and with fuel costs climbing, lower running and higher second-hand values give the GRP body a significant advantage.

U.K. and U.S. cars would still incorporate some steel. Although GRP is stronger than steel, weight-for-weight, more resistant to minor impact and more capable of absorbing impact up to its yield point, nevertheless steel has a higher flexural modulus and is capable of absorbing more energy after it has reached its yield point. By combining these characteristics it is argued that the result should provide better protection to personnel and less total damage to the car in the majority of accidents than a body built entirely of either GRP or steel. Another factor that contributes to increased safety is that the total weight of a four-door GRP saloon would be about 400 lb less than its steel counterpart thus the amount of energy to be absorbed in the event of an accident would be correspondingly lower.

8.1.2   Costs: tooling-up time

The cost of tooling up for the bodies is anticipated to be between 15–20% less than that for similar bodies in steel, assuming that the metal

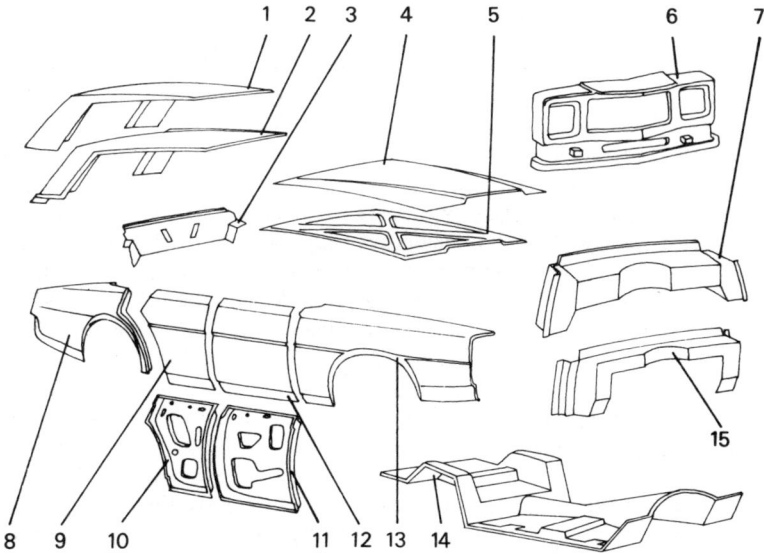

Fig. 8.1. Major GRP components of a 4-door saloon. (Courtesy Molded Fiber Glass Companies: Plastics Design Forum)

1. Roof panel; 2. Roof inner panel; 3. Rear seat bulkhead; 4. Bonnet panel outer; 5. Bonnet panel inner; 6. Radiator grille opening and headlight panel; 7. Dash panel; 8. Rear quarter panel (left and right); 9. Rear door panel outer (left and right); 10. Rear door inner panel (left and right); 11. Front door inner panel (left and right); 12. Front door panel outer (left and right); 13. Front wheel panel (left and right); 14. Undertray; 15. Plenum panel.

parts were available from other models. Additionally, the time required for tooling would be some 12 months shorter than would be required for a steel pressing line. With regard to production, the capacity for one complete set of tooling would be limited by the largest moulds and the slowest of the moulding operations. This would undoubtedly be the under-body which could be produced at a rate of about 44,000 per year. Smaller components of the body could be run at double this rate.

An idea of the mouldings involved can be gathered from Figure 8.1 which shows some of those used for a four-door saloon. The parts listed includes some which are not illustrated.

Figures quoted put moulding and assembly operations at optimum efficiency when production is at about 40,000 units per year, with no substantial economy being obtained if this number is exceeded. This estimate includes the use of an additional set of moulds and fixtures for producing the large underbody and the requisite number of moulds and fixtures for producing any of the other mouldings where the capacity of a single set would not be sufficient to handle the next operation in the production sequence.

Assembly and painting costs are difficult to predict. As an example, the above costs on the Avanti body built by Molded Fiber Glass Companies, Ohio, U.S.A., in 1963 were of the order of £100. While substantial developments have taken place since in coatings and coating techniques, labour costs have also risen considerably and today assembly and painting costs prior to the fixing of glass and interior trim would be considerably higher per body. It should be noted, however, that with the lead time between a decision to build and the production of a range of GRP bodies on the lines described, prices and labour cost no doubt will have risen substantially and thus the cost figures will have to be adjusted accordingly.

8.1.3   Production factors

In the quantity production of an all-plastics body the single most significant factor, weight reduction, compounds the designers' problems in that he has to marry the economics of the project with the ever increasing stringency of regulations governing safety. Hitherto the body designer did not have to worry unduly about weight reduction. Until recently this applied particularly in America where because of the availability of cheap fuel, bodies were designed primarily for visual appeal and comfort. In Europe the situation has been somewhat different. Fuel has always been relatively expensive, and for this reason and because of the different geography, leading to comparatively shorter journeys and pressure on parking space, private cars have been smaller and in the main far more economical with regard to fuel consumption. The difference still obtains, but not to the same extent. During the past three years General Motors, for example, have improved the efficiency of its range by some 50%. Understandably this advance has not been entirely due to size and weight reduction. Engineering has played its part, but as the figures stated earlier show plastics have made a substantial contribution to overall weight reduction and associated fuel economy. Today, because of legislation, the pressure to reduce weight is even greater and all the major companies are striving to be in a position to meet the mandated 27.5 m.p.g. figure by 1985.

The trend in America and possibly to a lesser extent in Europe will be towards smaller size, lighter weight and more aerodynamic body shapes. Current design projects aimed at integrating these factors point the way to some quite dramatic changes in body styles. As discussed in Section 8.2.1 a door design project carried out by Owens-

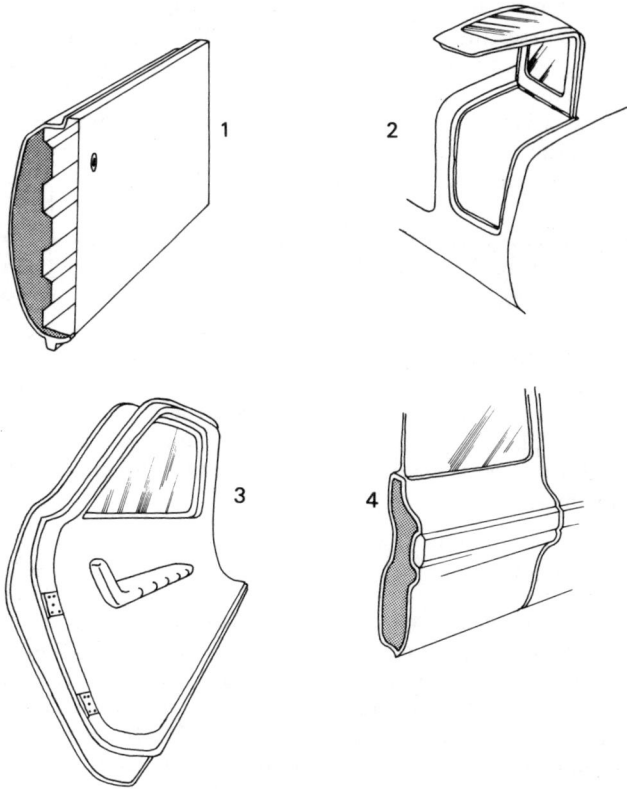

Fig. 8.2. Various door designs applicable to production in GRP. (Courtesy Owens-Corning Fiberglas Corporation)

Corning Fiberglass has examined a number of different door styles some of which are far removed from the conventional but are particularly suited to the use of GRP. What may well be termed 'futuristic' today will no doubt be looked upon as conventional in a few years time. Some of these designs are shown in Figure 8.2.

8.1.4 Weight analysis

To the designer of a moulded body the pressure to reduce weight means that every component must be analysed not only from the point of view of economies and ease of moulding but also from the service life angle. Where a component or panel is subject to low stress it can be moulded thin. Today, the necessity for using ribs or ·thickening can readily be calculated. To produce the assembly on a less scientific basis would almost certainly lead to economic failure due to increased material cost. This consideration is far more important in a GRP body where the materials are more expensive than steel.

In the design stage of a GRP body another important factor is the degree of knowledge required with regard to the materials themselves if the various elements of the design team are to carry out their work properly. Steel is a relatively simple material the characteristics of which are well understood. Plastics are a different matter and the necessary in-depth knowledge of their behaviour is not so readily available.

The reinforced materials with which we are dealing pose even more complex problems due to the wide range of resin and reinforcement combinations that can be used. Thus the 'knowledge' curve as applied to reinforced plastics, particularly in load bearing applications, is still steep and it will be some time before it assumes the same form that exists for metals.

8.1.5 Materials

In the type of project discussed reinforcement materials would consist of glass fibre mat or preform except where sheet moulding compound was specified. Exterior mouldings would use a low profile polyester resin and a 0.75 mm glass fibre overlay mat on the surface. Structural mouldings would be of isophthalic polyester. The resins used would not normally be flame resistant but mouldings could be produced in a flame resistant resin at slightly greater cost without difficulty.

Material specification is also closely associated with the safety factors set by the manufacturer and/or by legislation. In the U.S.A., major manufacturers are carrying out side impact tests on panels and doors in GRP, and are endeavouring to obtain results that match or exceed the requirements for steel doors currently in use. Methods vary, but in general the component is moulded to a predetermined specification but as light as possible within its limits. After tests, material is sometimes added to the component to overcome moulding problems caused by too thin a section or too rapid a section change which affect material flow in the mould. A weight saving of up to 40%, as compared with the steel counterpart, is typical.

8.1.6 Design iteration

In practice, to satisfy all the constraints it is a matter of preparing the design and then proceeding to stress analysis, back to design and then back again to stress analysis — termed 'design iteration'. The process is commenced during the initial design work with some basic analysis to determine the structure that best meets both the specification and the service requirements. After the initial drawings the iterative analysis technique is used, commencing with coarse analysis followed by successively finer steps terminating in a finite element analysis model. The process operates in two ways. In the first method, the designer will design a component or panel in plastics with sketches for the structural analyst who, in turn, will raise design suggestions based on a rough structural analysis. In the second method, the structural analyst will be given the panel at the commencement of the operation and the rough analysis will be turned over to the designer. In each case it involves design iterations. The advantage of using the structural analysis technique is that form and thickness can be modified to optimise material usage. Areas where more or less material is required can be examined by reference to stress plots. Position, size and configuration of ribs can be determined as well as wall thickness. Component integration can also be examined to ensure that, for example, a structural inner member is compatible with a visual outer panel. At the same time factors affecting production can be considered since the thickest part of a panel will determine moulding cycle time. Thus with this information to hand a decision to use, for example, a number of thin section ribs rather than one thick one may be seen to be advantageous with regard to the moulding cycle.

8.1.7 Computer analysis

One important technological advance which aids today's designers is the use of a computer. Originally plastics body components for mass production were designed on the same lines as steel stampings without due consideration of the greater flexibility of the plastics (Section 1.1) part. Thus while plastics add to the complexity of the designer's problems in obtaining rigidity their versatility can be used advantageously given sufficient data. It is here that the computer comes into the picture obviating the older, slower and less efficient techniques of building a prototype and attaching strain gauges. The computer permits the analyst to break down a model into small areas and to decide which can be reduced and which require additional stiffening. Briefly, the point at which a design project in GRP begins and ends is similar to that traditionally used for a steel body, the difference being that today the various activities involved during the process have changed.

8.2   GRP DOORS IN QUANTITY PRODUCTION: COST CONSIDERATIONS

As mentioned in the foregoing there is considerably development work being undertaken by the major U.S. manufacturers in the design of GRP doors. Unlike roof panels, for example, which do not involve hinges, sealing or side impact safety problems, automobile doors on a high throughput compression moulding basis present the designer with one of his greatest challenges. The following factors are taken into consideration: (1) production may be cost competitive with steel; (2) unit manufacturing costs may be either lower or marginally higher than its steel counterpart depending on design; (3) a high throughput process producing GRP would be liable to cost increase if output falls, investment costs rise or if there is an increase in labour rates. Other processes for door production, using aluminium, conventional SMC and reinforced thermoplastics have been found to be non-cost competitive and thus they would only be selected as a door production process for reasons other than cost. The one disadvantage of high throughput SMC is the relatively high raw material price which make it more vulnerable to material price increases.

However, the competitiveness of GRP from a production cost point of view in this case is not an advantage in itself but it puts the moulding process on an equal footing with steel. Its advantage is in a significantly lower capital investment. Calculation shows that a high through-put SMC moulding line on an equivalent capacity basis to a steel pressing line would require a 67% lower investment than the latter. Indeed a complete SMC line including tooling can be set-up for somewhat less than the capital outlay required for steel tooling — a significant factor considering that steel tooling is conventionally replaced every three years.

8.2.1   Capacity utilization: throughput sensitivity

The capital investment intensity of a pressed steel body line renders the cost structure vulnerable to any drop in production quantity. It will be appreciated that the cost of machinery and tooling is so great that tooling-up costs must be amortised over a very large production run in order to achieve a reasonable unit cost level. As can be seen from the sensitivity diagram for steel (Figure 8.3) at a one-shift level of five days/week, unit costs exceed $23.00: at a two-shift level unit costs fall to $18.00: while at three-shifts unit costs drop to $16.50. These figures show that a steel pressing line would have to be operated on a

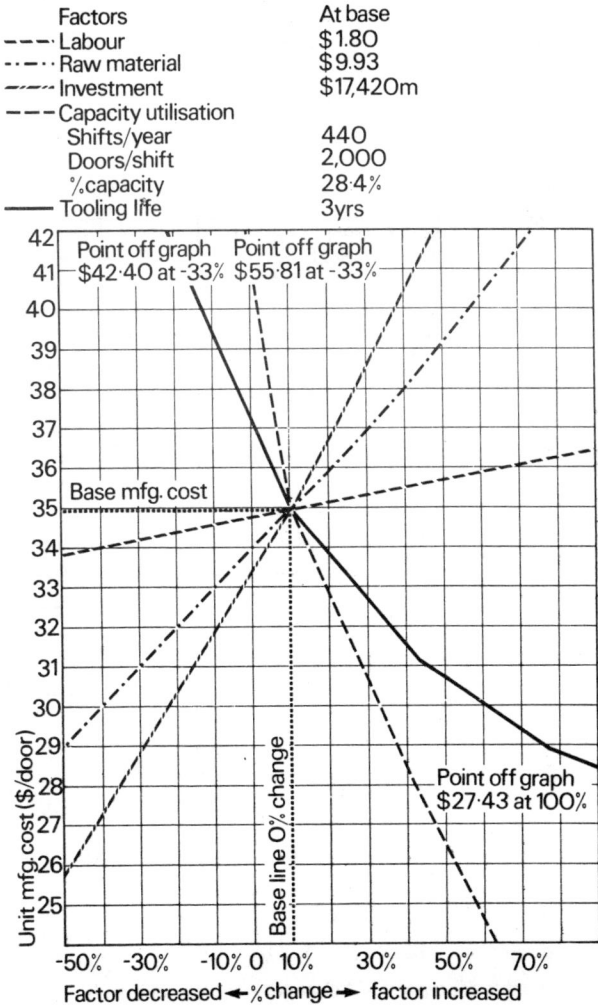

Factors	At base
--- Labour	$1.80
--- Raw material	$9.93
--- Investment	$17,420m
--- Capacity utilisation	
Shifts/year	440
Doors/shift	2,000
%capacity	28·4%
Tooling life	3yrs

Fig. 8.3. Sensitivity diagram showing unit costs for steel door construction on different shift bases. (Courtesy Owens-Corning Fiberglas Corporation)

three-shift level in order to obtain a lower unit cost than a high throughput SMC moulding line.

It is interesting to note that a SMC line operating on conventional lines would show a very similar cost pattern because it is as capital intensive as a steel pressing line on an adjusted capacity basis. The cost structure for aluminium, as previously mentioned, is very similar

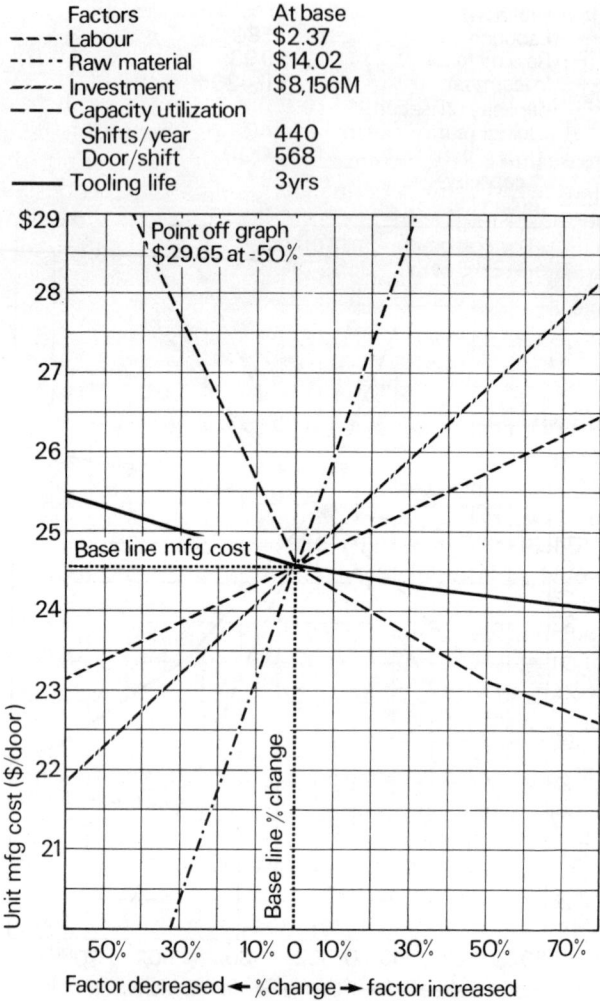

Factors	At base
--- Labour	$2.37
--- Raw material	$14.02
--- Investment	$8,156M
--- Capacity utilization	
Shifts/year	440
Door/shift	568
——— Tooling life	3yrs

Fig. 8.4.   *Sensitivity diagram showing unit costs for SMC door construction on different shift bases.* (*Courtesy Owens-Corning Fiberglas Corporation*)

to that of steel. However, it carries the additional burden of high material costs and thus unit costs are affected in the same manner as steel but at 70% of the production rate. SMC on a high throughput rate and GRP both have sufficiently low capital investment characteristics relative to throughput to be considerably less sensitive to capacity utilization. This is shown in Figure 8.4.

8.2.2 Cost sensitivity

Cost sensitivity studies have been carried out in the U.S.A.[21] to determine the effect of extended tool life on unit costs. It has been shown that a longer than normal tool life, or from four to six years, will decrease unit costs on a pressed steel door at a faster rate than its counterpart in SMC but the gain in terms of unit cost is not sufficient to exceed the cost advantage of SMC. Even if the tool life of a steel pressing line is raised to the maximum of six years and the tooling for SMC retained for only three years the latter would still show to advantage.

8.3 GRAPHITE FIBRE — COMPOSITE BODY COMPONENTS

While this book basically sets out to review uses of glass reinforced plastics (GRP) in automobile bodywork any reference to future trends would be incomplete without mention of advanced forms of reinforcement for resins. Among these developments applicable to automobile bodywork are graphite reinforced resins.

One of the most interesting potentially large volume application of the new composites is under development by Ford of Detroit, who are in the process of building a prototype car using, where possible, graphite fibre as a reinforcement for moulded panels and chassis frame[5]. Figure 8.5 shows the body panels which will be produced in graphite fibre reinforced resin, possibly polyester.

The reinforcement material being used in the project is polyacrylonitrile (PAN) fibres which are subjected to a high temperature process involving controlled burning of the fibres followed by heating in an air-free atmosphere. The result is a high strength, lightweight fibre which can be woven into a form of fabric or tape for combination with an epoxide or polyester resin. Three times lighter than steel, graphite fibre composites have specific strength, modulus, density, stiffness and coefficient of friction properties that are excellent compared with other materials.

8.3.1 Graphite composite prototype

The car being constructed will weigh some 2750 lb as compared with about 4000 lb if built conventionally, will seat six, and is expected to return a petrol consumption of not less than 23 m.p.g. under main

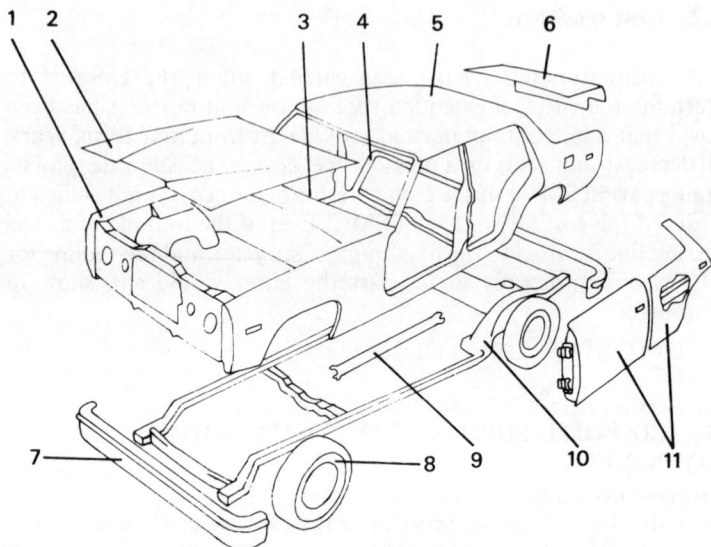

Fig. 8.5.   Graphite fibre/GRP panels for body project by Ford of Detroit. (Courtesy Plastics Design Forum)
1. *Front end assembly;* 2. *Bonnet;* 3. *Instrument panel and interior;* 4. *Front seat frame;* 5. *Body structure;* 6. *Boot lid;* 7. *Front and rear bumbers;* 8. *Wheels;* 9. *Drive shaft;* 10. *Chassis;* 11. *Doors, guard beam and hinges.*

road conditions. Powered by a V-6 engine of 2.8 litres capacity, it should accelerate from 0–60 m.p.h. in about 12 s.

The project will be accomplished in two phases. In the first, selected components in the new graphite composite will be built into existing Ford production cars. The second phase will involve designing and testing a 1979 car using as many graphite composite components as possible. The prototype is expected to be ready for evaluation early in 1979.

9

Quality control

9.1 DIAGNOSIS OF FAULTS

Many of the problems concerning the appearance and performance in service of GRP mouldings arise from undercure of the resin. There are other faults that occur in the form of both visible and sometimes invisible flaws and these are set out below. The remedy in each case becomes apparent from an analysis of the cause. The following concerns components produced by methods including SMC moulding.

9.1.1 Delamination of gel coat

This fault (Figure 9.1) can be difficult to detect unless very pronounced and the gel coat flakes when the moulding is handled in finishing operations. Sometimes areas of poor adhesion can be detected by the presence of a blister or by local undulations when the moulding is viewed obliquely. The cause of poor gel coat adhesion is due to inadequate consolidation of the laminate or by the fact that the gel coat has been left to cure for too long. Another cause can be the use of soft mould release waxes which can penetrate the gel coat and adversely affect adhesion.

9.1.2 Surface wrinkling

This unfortunately quite common fault (Figure 9.2) is caused by solvent attack from the monomer in the laminating resin and is due to undercure of the gel coat. Surface wrinkling can be obviated by making sure that the formulation is correct and that is thickness is adequate and consistent. Other factors which can cause the problem are incorrect shop temperature and/or humidity and sometimes the fact that the gel coated mould has been in a draught. For the latter

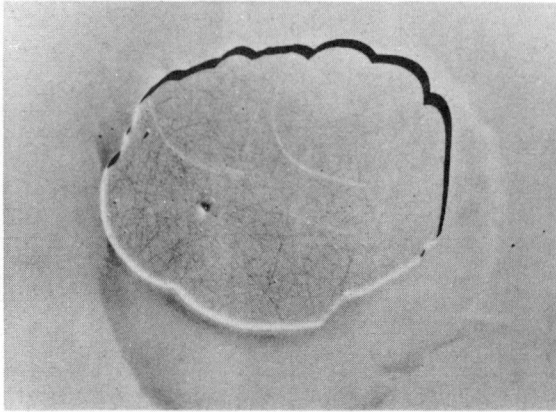

Fig. 9.1.   *Poor gel coat adhesion*

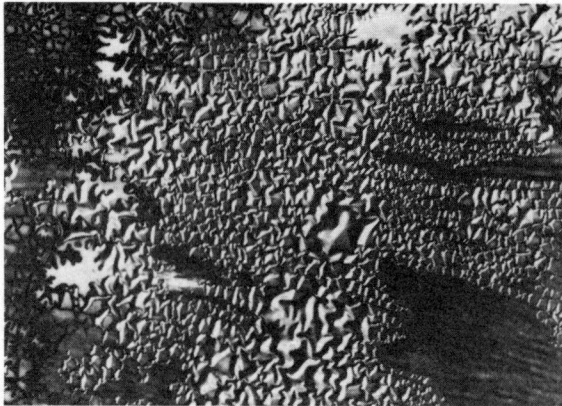

Fig. 9.2.   *Surface wrinkling*

reason gel coats should not be accelerated by directing hot air blowers directly into the mould. The application of a second gel coat before the first has gelled sufficiently can lead to softening and interaction between the two causing wrinkling.

9.1.3   Blistering

The presence of blisters on the surface indicates delamination of the gel coat, as described, or that there is delamination within the

Fig. 9.3. Pinholding (microphotograph)

laminate and that air or possibly solvent has become entrapped. If the blisters extend over a considerable area the indications are that the resin is undercured. This type of blistering may not occur until some months after the moulding has been produced. Another cause of this type of blistering is the excessive use of radiant heat lamps to accelerate cure. The use of an unsuitable grade of MEKP catalyst can also produce extensive surface blistering.

However, if the blistering is below the surface of the moulding the most likely cause is imperfect wetting out of the reinforcement during lamination. This would be due to insufficient time being allowed for the reinforcement to absorb the resin before rolling. Apply a generous layer of resin over the gel coat so that when laying-up air is forced up through the mat. Blisters of this type can usually be detected by a close inspection when the moulding has been removed from the mould.

9.1.4 Pinholing

This is another not uncommon fault (Figure 9.3) and is caused by small air bubbles which become trapped in the gel coat prior to gellation. The trouble arises if the gel coat resin is too viscous, has too high a filler content or if the gel coat fails to wet the release agent on the mould surface properly. Other causes can be traced to vigorous stirring of the gel coat resin immediately prior to application or the presence of dust particles in the resin, brush or in both. When brush applied gel coats should be laid with even strokes. Avoid using a stippling action.

Fig. 9.4.   Crazing (microphotograph)

9.1.5   Crazing

Crazing of the gel coat (Figure 9.4) can occur quite soon after removal of the component from the mould or may take some months to appear. The fault appears as fine hair cracks in the resin surface. Sometimes the cracks are so fine that the only indication of the trouble is that the moulding appears to have lost its surface gloss. Usually crazing occurs in resin rich areas and is caused by the use of an unsuitable resin formulation for the gel coat. The addition of an excess of styrene in the gel coat is a common cause. Alternatively, the gel coat resin may be too hard relative to its thickness. Where thick gel coats are required a more resilient resin should be selected. Where crazing only appears some months after moulding and exposure to weathering or to chemical attack it is most probably caused either by undercure, the addition of an excess quantity of filler, or the use of a resin that is too flexible.

9.1.6   Star crazing

This problem (Figure 9.5) should not be confused with crazing over the surface of the moulding as it is the result of using a thick gel coat and occurs when the laminate has received an impact on the underside. Gel coats should thus not be thicker than about 0.4 mm.

9.1.7   Internal dry areas

These areas in a moulding are caused by an attempt on the part of the operator to impregnate more than one layer of mat at a time. Dry or

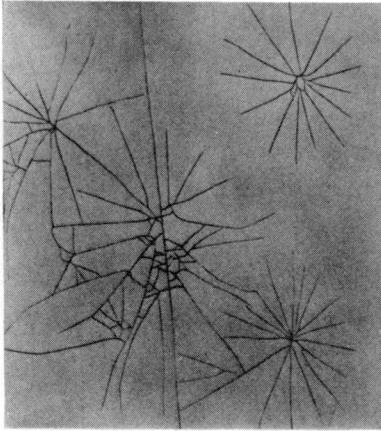

Fig. 9.5. Star crazing

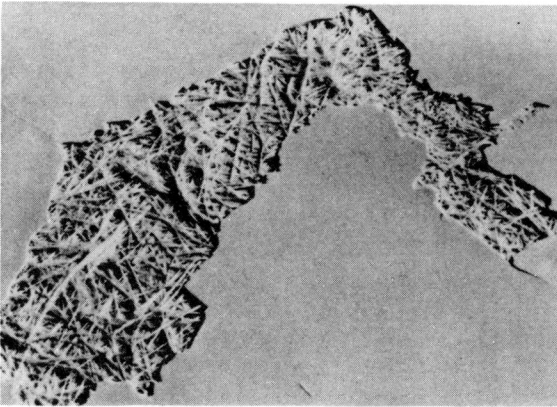

Fig. 9.6. Internal dry patch

imperfectly impregnated areas, Figure 9.6, can be confirmed by tapping the surface of the laminate with the edge of a coin or similar light object.

9.1.8 Imperfect wetting out of reinforcement

This defect is normally only apparent on the underside of the moulding where there is no gel coat. When properly wetted out this

Fig. 9.7.   Severe leaching

surface will present a somewhat glazed appearance due to all the glass fibres being covered with resin. The cause of dryness is either the use of insufficient resin during lay-up or inadequate rolling and con- solidation aimed at bringing the resin right through the reinforcement to the surface of the under side. Dryness at the top edges in a deep mould can be traced to imperfect consolidation and in some cases to draining of the resin away from the edges. The use of a thixotropic resin on vertical surface moulds cures the trouble.

9.1.9   Leaching

Leaching (Figure 9.7) is a serious fault and occurs after exposure of a moulding to the weather. It is characterised by loss of resin from the laminate which leaves the glass fibres exposed to attack by moisture. The cause is that the resin has not been properly cured or that the choice of resin for the application is incorrect.

9.1.10   Spotting

This fault takes the form of small spots over the entire gel coat surface of the moulding and is most commonly caused by inadequate mixing or dispersion of one of the constituents of the gel coat resin.

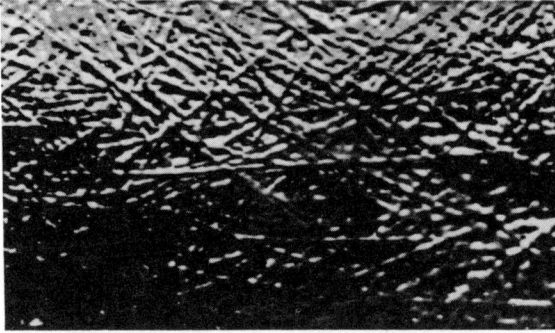

Fig. 9.8.   Fibre pattern

9.1.11   Striations

This effect is normally caused by pigment floatation and is most likely to occur where the colour used is a mixture of more than one pigment. The remedy is very thorough mixing of the pigments or the use of a different pigment paste.

9.1.12   Fibre pattern shows

Sometimes the pattern of the reinforcement is visible through the gel coat or can be seen on its surface, Figure 4.8. The problem usually occurs if the gel coat is too thin or if the reinforcement has been laid-up and consolidated before the gel coat has cured sufficiently. It can also happen if the moulding is removed too soon from the mould.

9.1.13   'Fish eyes'

This phenomenon (Figure 9.9) occurs sometimes when a very highly polished mould is in use and in particular when a silicone modified wax is used for release. In these cases the gel coat can tend to 'de-wet' from certain areas leaving spots where the gel coat is almost non-existent. Where this occurs pale coloured patches, usually about 6 mm in diameter, can be seen. A similar effect can also occur in long straight lines which follow the brush strokes used. The fault rarely occurs when a mould release film is correctly applied.

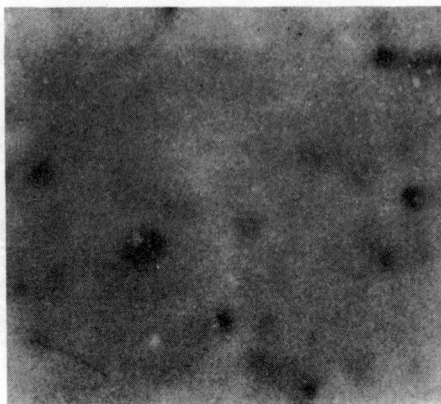

Fig. 9.9. 'Fish eyes'

9.1.14 Yellowing

This problem rarely arises in automobile work as the laminate is painted. On unpainted laminate yellowing occurs after a period of exposure to sunlight. Usually the effect is only slight and is due to the absorption of UV rays. It does not affect the mechanical properties of the laminate, but those with a high glass content tend to discolour more rapidly than a resin rich moulding.

9.1.15 Resin not set in patches

Not uncommon is the fault which exhibits itself in the form of white patches of ungelled resin. This can usually be traced to dampness in the mould, the presence of dampness in the brushes used to apply the laminating resin or imperfect dispersion of catalyst. Other reasons which can lead to this problem are the presence of undried PVA release agent on the surface of the mould or the use of damp reinforcement.

9.1.16 Sticky laminate surface

The problem of a laminate surface which remains sticky after curing is usually caused by insufficient catalyst or accelerator in the resin to ensure complete cure, excessive evaporation of styrene (see Section 9.1.2) or the use of a damp mould. The effect can also occur through the use of an unsuitable parting agent or through the surface of the mould being imperfectly sealed. This fault will also produce problems in demoulding with the component tending to stick in the mould.

9.1.17 Excessive exotherm

Excessive heating up of the laminate during cure can cause discolouration and in some cases can lead to distortion of the mould. At best it will tend to destroy the film of wax release agent on the mould surface. The problem can be avoided by attention to the catalyst and accelerator ratios and by reducing the thickness of laminate laid-up at one time. Use lighter chopped-strand mat and more layers.

9.2 FAULTS OCCURRING IN SMC MOULDINGS

Because of the different nature of SMC and GRP laminate produced with a gel coat either by hand spray-up or other methods the problems associated with SMC arise for different reasons. The following is intended only as a guide to identify and correct the more common types of defect encountered in SMC moulding. It will be appreciated that situations will occur that require a thorough study in order to achieve success and the most that any written guide can hope to achieve is to suggest avenues of investigation.

9.2.1 Blistering

In SMC moulding blistering is due to air or gas trapped between layers of the material. It can be overcome by a reduction in the size of the charge, by an increase of material charge at the flow centre and/or a reduction in moulding temperature.

9.2.2 Porosity in or near thick sections

This is caused in or adjacent to thick sections by entrapment of air. The solution is to increase the material charge in heavy areas, to relocate the charge in the mould with the object of sweeping the air out of ribs, etc. or to reduce the charge area.

9.2.3   Random porosity

This type of porosity can be caused by material pressures being too low, in which case an increase in the charge weight or an increase in moulding pressure should cure the problem. Porosity can also be caused by worn shear edges. This fault can be overcome by an increase in the charge weight, an increase in punch temperature and/or a reduction in cavity temperature. Obviously the most effective solution is to repair the tooling.

9.2.4   'Short' mouldings

'Short' or incomplete mouldings can occur through insufficient weight of SMC in the mould, insufficient flow in the material under pressure or the application of insufficient moulding pressure. The first cause can obviously be overcome by increasing the charge weight while poor flow can be cured by increase in pressure, increase in charge weight or a reduction in temperature. Other causes of 'shorts' can be due to improper location of the charge in the mould which can be cured by relocation to the non-fill area or can be due to the use of SMC which is outside its storage life.

9.2.5   Distortion

There are a number of causes of distortion in an SMC moulding. Fibre orientation is one which can be eliminated by adjusting the charge size for minimum flow. Excessive shrinkage can induce distortion. It can usually be cured by changing the grade of SMC used for one that gives lower shrinkage. Handling can cause distortion and ejector systems and operator's methods of demoulding should be reviewed. Any stress conditions during cooling or finishing should be eliminated. Distortion can also be attributed to the design of the plant and the use of a cooling fixture and/or water bath can be advantageous. Undercure is also often the cause of distortion and here the simple solution is to experiment with increased cure time.

9.2.6   Poor release: sticking in the mould

This fault can also be due to a number of causes which include too low a mould temperature, a poor mould surface, possibly with undercuts, porous areas or heavy machining marks. The use of an external release agent in the mould can often overcome the problem. Improper design of an ejector system can cause difficult release. All ejectors should be checked for clean and efficient action, and ejectors added in rib areas to prevent keying. Dirty tools can be cleaned and moulding surfaces coated with a suitable release agent. Excessive shrinkage can be a cause of sticking in certain shaped moulds. A possible cure is to use a grade of SMC with different shrinkage; provide undercuts in vertical tool surfaces to hold the moulding on to the desired half of the tool; check draft angles on all vertical surfaces.

9.2.7 Cracking of mouldings

Cracks are caused largely by excessive stressing (ejector systems should be checked) through fibre orientation, which can be cured by adjusting the charge size and location for minimum orientation and flow, and through the excessive build-up of exotherm in thick sections. In the latter case this will mean redesigning the mould to minimise thick sections.

Cracking can also be due to flow weld lines caused by flow of material from two directions and not joining up properly. It can be overcome by adjusting the location of the charge to minimise material flow around obstructions in the mould or by adding a heavier section to the area below the mould restriction to create turbulence and eliminate weld lines. Check that mould pinch-off clearance is uniform; consider the use of an alternative grade of SMC.

9.2.8 Poor surface gloss

This problem is usually due to undercure or to the use of too low a moulding pressure.

9.2.9 'Dieseling'

'Dieseling' is a fault caused by the compression of air/styrene vapour within the mould cavity which causes the mixture to ignite and locally blacken the moulding. The cure is to arrange the charge in such a manner that all the air is expelled as the material flows within the mould.

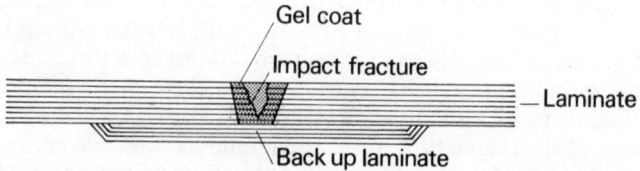

Fig. 9.10. *Repair of a punctured moulding. (Figs. 9.10–9.12 by courtesy of British Industrial Plastics Ltd.)*

9.2.10 Shrinkage marks

These are caused by uneven shrinkage of the moulding during the curing period. The cure is to increase the weight of the material and/or remove the pressure pads which limit the downward travel of the upper part of the mould.

9.3 REPAIRS TO GRP MOULDINGS AND BODY PANELS

Mouldings produced by any of the moulding methods other than SMC or DMC can be repaired, unless very badly damaged, in a comparatively simple manner. Repairs to be carried out in the moulding shop itself can be dealt with at the trimming stage with the advantage that the resin will not have matured, and thus will facilitate a good bond. For superficial damage, for example, to the gel coat only activated resin with the addition of a small percentage of thixotropic agent should be applied and allowed to gel. In some instances, covering with a sheet of cellophane will help to maintain the resin in position and will give a smooth finish. However, it is usually advisable to apply somewhat more resin than necessary to fill the repair in order to allow for shrinkage and flatting off of the moulding when cure has taken place in which case the smooth finish due to the cellophane is of no great advantage.

9.3.1 Impact damage

The repair of more extensive damage such as impact fractures which usually take the form of a crack extending right through to the underside of the laminate, must be carried out in stages.

(1) The fracture is enlarged from the surface which sustained the impact to form a V-shaped groove ensuring that all damaged laminate is removed through the full depth of the crack, Figure 9.10.

(2) The underside of the laminate is sanded to a point not less than

Metal plate   Cut away and filled

Gel coat

Laminate

Back up laminate

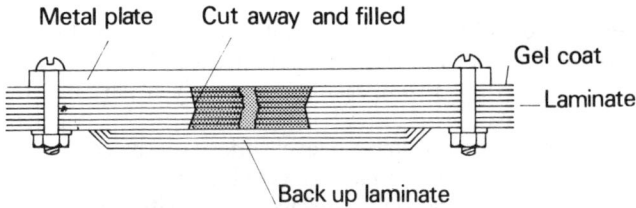

Fig. 9.11. Repair of large impact damage

about 75 mm in all directions from the groove. All loose strands and powder are brushed away.

(3) Four strips of chopped strand mat (450 g/m^2 is usually suitable) are prepared. The strips should be of increasing size, the smallest extending about 25 mm beyond the edges of the prepared slot with the remainder being progressively larger by about 12 mm all round.

(4) The strips of reinforcement are weighed and approximately $2\frac{1}{2}$ times their weight of resin is catalysed and accelerated.

(5) Working back from the laminate, a liberal cost of resin is brushed over the entire sanded area. The smallest strip of mat is laid centrally over the slot, more resin is applied taking care not to push the mat into the slot and rolled until impregnation is complete. The remainder of the strips are applied in a similar manner, finishing with the largest.

(6) The repair is allowed to cure, if possible being left overnight.

(7) When cure is complete the crack is progressively filled with resin impregnated pieces of mat to within about 2 mm of the original surface.

(8) Lastly, a suitably filled and pigmented mix is trowled into the crack until just proud of the surface. When fully cured this can be rubbed down until level with the original gel coat.

9.3.2 Punctured mouldings

In cases where the moulding has been more severely damaged and an appreciable hole has been made the following procedure should be adopted. Figure 9.11.

(1) All rough edges and fractured laminate are cut back to produce smooth edges all round. The underside of the laminate is sanded, as before, in the surrounding area.

(2) As shown in Figure 9.11, the edges of the hole are rasped to a wedge shape from both sides. Thin laminates can be chamfered from the underside only.

(3) Sections of chopped strand mat, if possible of a thickness equal to that of the original laminate, are cut to fit the hole and larger pieces to be used as backing are cut, each progressively larger up to the size of the sanded area.

(4) In order to obtain a smooth level surface on the gel coat side of the repair, it is necessary to attach a sheet aluminium or other suitable rigid material on to the outer surface either with clamps or with strong adhesive tape. If the repair is large and is to be made on a flat area of the moulding it is preferable to bolt the metal sheet to the laminate. The bolts should be as close as possible to the hole to be repaired but must be clear of the area later to be covered by the backing laminate.

(5) Before being placed in position the inner face of the metal plate should be generously coated with wax and given a good coat of release agent preferably polyvinyl alcohol.

(6) When the release agent is quite dry the inner surface of the plate is given a heavy layer of filled and, if necessary, primented gel coat resin working through the hole with a brush.

(7) When touch dry a generous layer of catalysed and accelerated laminating resin is applied to the gel coat followed by successive layers of chopped strand mat. As before, each layer is thoroughly impregnated and rolled to remove any air bubbles. When the hole has been completely filled flush with the underside of the moulding the laminate is left to cure. It should be noted that when very large holes in thick mouldings are being filled any great build-up of laminate can produce excessive exotherm at it cures and in these cases work should be suspended half-way and recommenced when the first few layers have semi-cured and any exotherm has dissipated.

(8) The plate is removed and the bolt holes filled with a catalysed resin/filler.

(9) As a final operation the backing laminate is applied with each successive layer being impregnated and rolled as before. Any low areas can be filled with pigmented resin, flatted off and polished, or painted if the repair is being carried out on a fitted body.

The above procedure applies to a flat moulding as regards the plate. If a large hole is to be repaired in a curved panel, a plate of the exact contour can be made in GRP either from the original mould or from an undamaged replica of the panel.

9.3.3 Repairing 'blind' panels

There are instances where it is not possible to have free access to the underside of a damaged panel, for example, in double skin con-

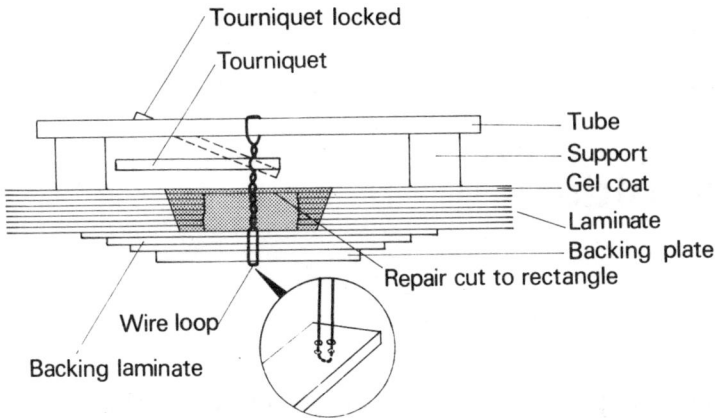

Fig. 9.12.   Repair of a 'blind' panel

struction. In these cases a somewhat different repair technique is adopted. The approach is illustrated in Figure 9.12.

(1) The first operation is to cut away the entire damaged area in the form of a rectangular hole. The underside is sanded as far as possible working through the hole.

(2) The edges of the laminate are filed taper as shown.

(3) A rectangular plate is prepared from plywood or metal. This plate has to be passed through the rectangular hole, thus its smaller dimension must be slightly less than the larger dimension of the hole. Two small holes are drilled in the centre of the plate to accept a pull wire.

(4) Rectangles of mat are impregnated with resin and laid on the plate which is then passed through the hole.

(5) Whilst the laminate on the plate is still wet the plate is pulled back using the wire loop to press the laminate against the underside of the panel.

(6) When the backing laminate has cured, the wire is cut back flush and the hole filled with impregnated mat until level with the upper surface. A final layer of gel coat resin can then be applied followed by a layer of cellophane which can be left until cure is complete. Finishing operations can follow those previously described.

10

Moulds

10.1 TYPES OF MOULD

As will have been appreciated from Chapter 1 moulds are basically of
two types; (1) 'open' type moulds such as used for hand and spray lay-
up methods of moulding, which in practice include moulds for
systems using vacuum and pressure bag moulding, and (2) 'closed'
moulds, usually referred to as 'matched', in which there is a male and
female half, both of which are rigid.

Materials of construction vary and in the main are dependent on
whether the system of moulding operates at low or high pressure. The
length of the production run will also be a factor in the choice of
mould material as to a lesser extent will be the degree of surface finish
required on the component.

Open moulds for hand and spray-up and other systems mentioned
(Sections 1.6.1 and 1.6.2) are usually of polyester/glass or epoxide
glass as the pressures involved are not great. Production moulds for
hot-press, SMC and DMC moulding are almost always of metal. For
short to medium length production runs the material can be
aluminium or are of the castable alloys such as Kirksite. For long run
work moulds for these systems are of steel with hard chrome
moulding surfaces.

10.2 OPEN MOULDS

Open moulds for hand and spray lay-up can be male or female
depending upon which side is required to have a mould finish. Male
moulds will produce a good finish on the inside of a shaped
component such as a facia board glove box, whereas a female mould,
by far the most widely used in automobile work, will produce a mould
finish on the outer surface.

Methods of producing these moulds are similar irrespective of type.
As mentioned briefly in Section 4.2, work is commenced for a female

mould with the construction of a full size master pattern. This can be made from a number of materials, the choice being dependent upon component size, complexity and the expected production life of the pattern. For producing one-off or a very limited number of moulds the pattern can be of plaster built-up on a hollow armature and finished with a hard and pore-free surface of resin, shellac or similar surfacing material. An alternative is to produce one female mould and to use this to produce a master pattern from which to mould further female moulds should duplication be required to meet increased production or to replace a mould which has become worn or damaged.

10.3 MASTER PATTERN: FEMALE MOULDS

The master pattern is an exact replica of the finished component and for a large moulding or indeed a complete body serves as the starting point from which all open-type moulds are made. Consider, for example, the pattern for a complete body to be produced by hand, spray-up or one of the low pressure systems. The first stage is either the construction of an accurate small scale model from which the full-size master can be lofted or as practiced by most manufacturers, the building of an accurate full scale model from the outset.

Methods of working from a small scale model can vary according to the sophistication of the personnel and equipment available. For the amateur or small builder the method used by the author is practical and requires the very minimum of equipment. The complete body model including items such as wheels and bumpers is produced in plasticene to a 1:12 scale to check styling, seating, engine space and general balance. From the model a plaster of Paris mould is cast, split to permit undercuts to be withdrawn. After a certain amount of careful internal smoothing, the mould is waxed, reassembled and a GRP moulding laid-up using firstly surface tissue followed by layers of the lightest chopped strand mat. When cured the GRP model is removed from the piece-part mould and lightly sanded. To obtain the main loft lines for the full size pattern, the model is scribed-off at equal stations using the front axle centreline as a datum and then carefully cut into sections at each station. These sections, which give the contours of the body at each station, are transferred to squared paper from which it is a comparatively simple matter to translate them full size to plywood templates. Working from a central datum line on each the templates are set-up on a rigid baseplate and made firm with timber stiffening (Figure 10.1). Fine mesh wire netting stapled between each template forms an armature for subsequent plastering

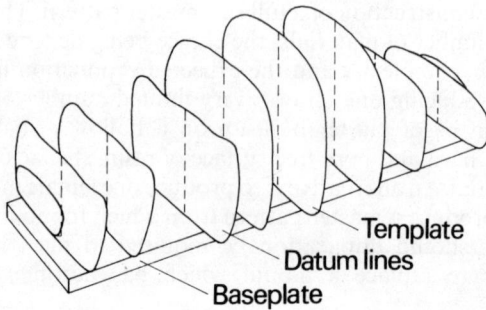

Fig. 10.1. *Wooden template set-up to datum lines*

and drawing out of the body lines to form a hollow but stable master pattern. When quite hard the surface of the plaster is sanded, filled with successive coats of shellac and, as already described in Sections 4.1 and 4.2, finally given a generous but smooth coat of PVA release agent. An alternative method of finishing a master pattern is to coat it with a furane resin. This cures hard and can be given a high and permanent surface suitable for the production of more than one female mould should they be required.

10.4 STYLING MODEL: QUANTITY MANUFACTURE

The large automobile manufacturer with styling and model building departments commences a new body design by far more sophisticated methods. Work begins by constructing a 1:4 or 1:8 scale clay model produced to visuals originated by the stylists. When the model is accepted a full-size clay model is built using the armature principle to reduce the quantity of clay required and the weight of the structure. Accuracy of contour is ensured by the use of templates originated from skin line drawings. The visual aspect is optimised by the use of thin aluminium foil to simulate stainless and chromium plated components.

10.5 LAYING-UP A PIECE-PART MOULD

Where undercuts occur in the master pattern it is necessary to split the female mould in order to be able to extract a one-piece moulding when complete. Thus before commencing the lay-up of the mould a careful appreciation should be made of the contours of the body in areas such as door return openings, boot and bonnet openings, where

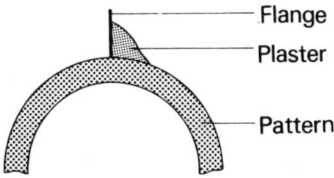

Fig. 10.2. *Section through the flange on a piece-part mould showing the 'fence' set-up prior to lay-up*

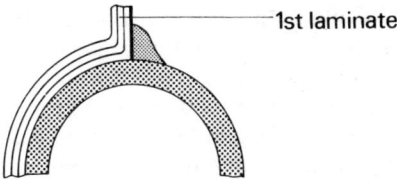

Fig. 10.3. *Lay-up on one side of 'fence' is complete*

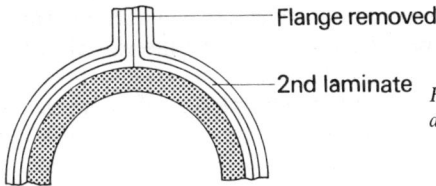

Fig. 10.4. *The 'fence' is removed and the second side completed*

weather channels are moulded-in, and at areas of return curvature. Where mould splits are required, for example, on the crown of a wing, the practice is to form a moulding edge or flange in plywood, hardboard or rigid plastic sheet (Figure 10.2) up to which the glass reinforcement can be laid. The flange when positioned is held in place with a backing of plaster. Before lay-up the face of the flange is well coated with parting agent to ensure clean release but is not gel coated.

During lay-up the laminate is brought up the flange as each successive layer of reinforcement is added, impregnated and rolled out (Figure 10.3). Particular care is taken to avoid air bubbles beneath the gel coat and thus a surfacing mat should constitute the first layer of reinforcement. Moulds which are to be used for extensive production runs are made with epoxide/glass rather than polyester/glass. The epoxide gives a harder and longer wearing surface. The method of laying-up is similar.

When the first section of the mould has cured, the flange is removed, the surface of the pattern cleaned of plaster and again waxed and coated with parting agent. The laid-up face of the first section is treated in the same way. Lay-up of the second section is then carried out in a similar manner to the first, Figure 10.4.

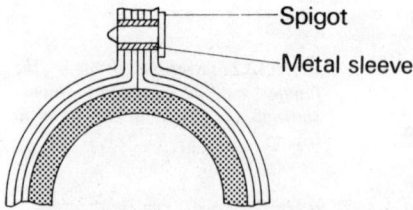

Fig. 10.5. *Spigot holes for location of the mould halves*

As stability is an important consideration in mould construction, the thickness of the laminate should be greater than that of the moulding it is to produce. Small moulds, if about twice component thickness, will not require stiffening. For large moulds some form of stabilizing structure will be required with the addition of a rigid base or if of an awkward shape, from the point of view of the laminator, it will need to be mounted on a trunnion pivotted frame. Stiffening can take the form of top-hat sections overlaid with random fibre mat or woven roving. Mounting points should cover a reasonably wide area and are usually of multi-ply wood or can be fabricated using reinforced plastics tubing.

All stiffening and mountings should be completed before the piece-part mould is removed from the master pattern. The final operation before removal is the drilling of the flanges for locating spigots and clamp bolts. For short run moulds it is only necessary to drill fitting holes to accept bolts with washers each side. Where moulds are to be assembled and disassembled regularly it is practice to mould-in metal sleeves to accept close fitting spigots in order to ensure alignment of the mould havles (Figure 10.5). After removal, the mould is assembled polished out, thoroughly waxed and prepared with parting agent as explained earlier. Waxing the assembled mould will tend to fill the mould parting lines but inevitably the join will be discernable on the moulding and will require light sanding. This fact should be considered when locating the mould split lines.

For quantity production, manufacturers use part moulds, often made of wood, taken from the master. Figure 10.6 shows a wooden master made by Pressed Steel Fisher, BL Components Ltd. Figures 10.7, 10.8 and 10.9 show plaster moulds for experimental door inner panels and main body shell component. Figure 10.10 shows the wooden framing used to support the latter mould. Figure 10.11 shows a plaster mould for a facia panel.

10.6 EPOXIDE/GLASS PATTERNS AND MOULDS

Long life master patterns and moulds are often produced using an epoxide resin in place of polyester. The system of laying-up both is

Fig. 10.6. *Positive wooden master model for saloon body in GRP. (Courtesy Pressed Steel Fisher BL Components Ltd)*

Fig. 10.7. *Plaster mould for door inner panels. (Courtesy Pressed Steel Fisher BL Components Ltd)*

Fig. 10.8. *Plaster mould showing divisions for saloon main body shell.* (*Courtesy Pressed Steel Fisher BL Components Ltd*)

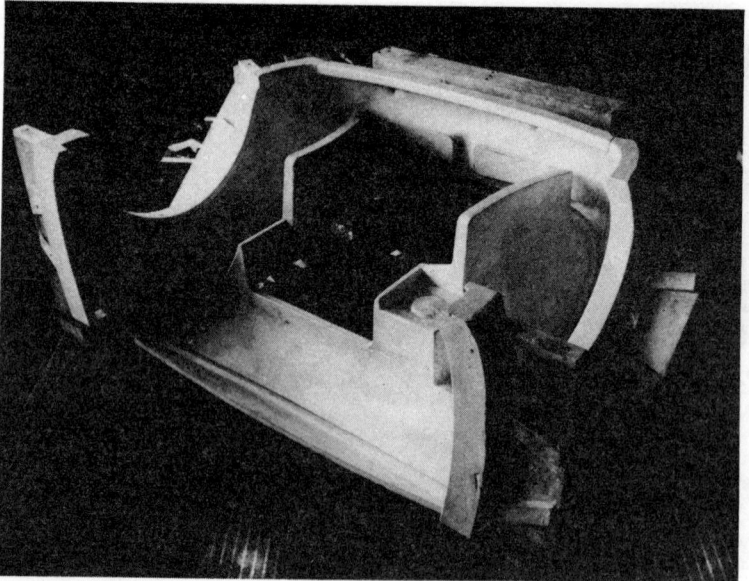

Fig. 10.9. *Plaster mould for front body unit.* (*Courtesy Pressed Steel Fisher BL Components Ltd*)

Fig. 10.10. Plaster mould showing wooden framing. (Courtesy Pressed Steel Fisher BL Components Ltd)

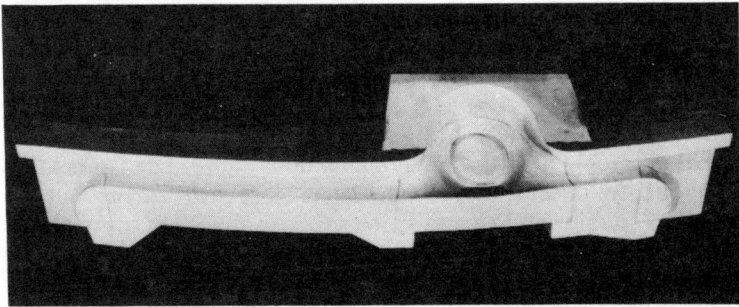

Fig. 10.11. Plaster mould for facia panel. (Courtesy Pressed Steel Fisher BLComponents Ltd)

somewhat similar but in many instances requires a greater degree of skill. The system recommended by Ciba–Geigy, using the company's Araldite LV569 or LV572 resin impregnated glass and hardener, uses wooden master patterns with epoxide moulds which must be given an efficient moisture barrier by treatment with a cellulose, polyurethane or furane varnish. Shellac should not be used.

The first operation is to apply a 0.75–1.5 mm gel coat which is brushed on. Three types of gel coat are available: Araldite SV411 with hardener HV411 is a black iron filled formulation, supplied pre-weighed: Araldite SW404 with hardener HY404 is a blue silicon carbide filled gel coat, also preweighed: Araldite SW2402 used with hardener HY2402 is available in a range of four colour-stable formulations differing in colour only.

Araldite SW2402 gel-coats are suitable for laminate structures designed to operate at temperatures up to 50°C. The Araldite SW404 gel-coat is suitable for structures operating up to 75°C; this improved heat resistance enables it to be used in combination with Araldite LV572 + HY560. The Araldite SV411 gel-coat offers the best heat resistance: its use with Araldite LV572 + HY560 permits an increase in working temperature to a maximum for the laminates of 80–85°C. Laminates of Araldite LV572 + HY561 require the use of the Araldite SV411 gel-coat since these laminates are normally designed for service at tmeperatures up to about 100°C.

When the gel-coat is almost tack-free (approximately 1–2 hours depending on temperature) lamination is applied in the form of chopped strand mat, scrim or cloth about 15×15 cm to 25×25 cm square. If the mould has awkward corners these are first filled with a tooling grade cotton flock or short glass fibre and resin infill mix. Further layers of reinforcement are laid, unlike polyester work, each being impregnated with the resin before being applied. As each layer is applied consolidation is ensured by thorough rolling with a washer type tool, Figures 10.12, 10.13 and 10.14.

When lamination is complete the mould can be stiffened by the application of lengths of 'Primatube' epoxy/glass fibre reinforced tube, which is first gel coated, laid on the back of the laminate using Araldite rapid adhesive or Araldite putty. The tubes are then overlaid with the required number of mat laminations impregnated as before. On moulds of difficult shape, stiffening is more easily carried out by the use of spirally cut Primatube. This tube is somewhat springy and thus over curved surfaces are normally held down in position with sand-filled polyethylene bags until the adhesive has cured.

10.7 MOULD LAY-UP WITH ALUMINIUM HONEYCOMB

An alternative method of stiffening large, generally flat moulds is to use 'Aeroweb', an aluminium honeycomb. The use of this material to form a lightweight, composite core greatly reduces the time required for the complete lay-up. With only two or three plys overlaying the honeycomb it produces a balanced laminate structure. The procedure is as follows: The Aeroweb aluminium honeycomb is cut to size using

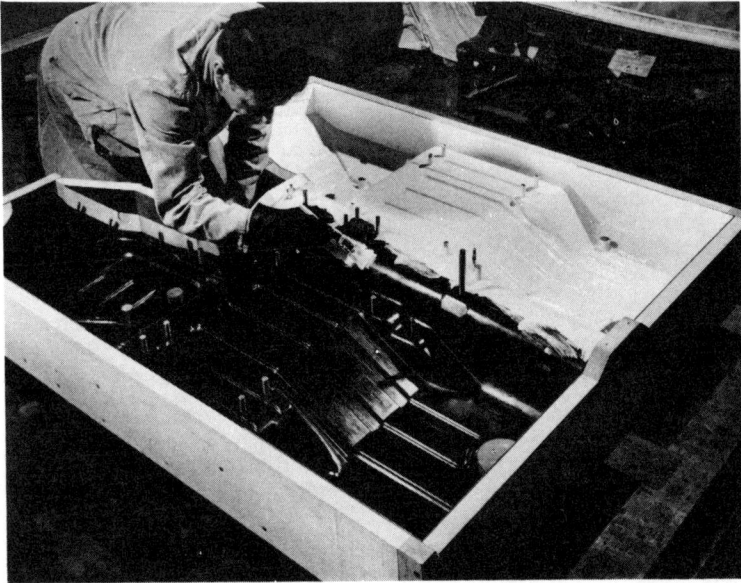

Fig. 10.12. *Araldite negative master model for a car main floor pan. (Courtesy CIBA (ARL) Ltd)*

Fig. 10.13. *Construction of main floor pan. Laying-up the first plies of glass scrim and Araldite resin. (Courtesy CIBA (AFL) Ltd)*

Fig. 10.14.   *Applying resin impregnated glass mat to the floor pan during the build-up of the laminate. (Courtesy CIBA (ARL) Ltd)*

a stiff steel knife (various thicknesses are available depending on the size of the mould). The master pattern is polished and pre-treated as before, followed by a gel coat and, when almost tack-free, the first lamination of surfacing mat is applied. Lamination is continued until two or three layers are built-up. Using a soft roller the underside of the honeycomb is primed with resin and the honeycomb sheet located on the laminate with the primed surface in contact with it. Again, using the soft roller the upper surface of the honeycomb is primed and an identical laminate laid up. As when applying Primatube stiffening tubes, it may be necessary to use sand-filled polyethylene bags to stop the honeycomb springing during cure. As a final operation the edges of the honeycomb are wrapped with resin impregnated glass fibre or cloth.

Where required, high strength areas can be built into the honeycomb sheet where attachments are to be made by infilling the honeycomb in these places with Araldite putty or modelling paste before overlaying the honeycomb core with laminate.

10.8   MATCHED GRP MOULDS: LOW PRESSURE SYSTEMS

Polyester/glass and epoxide/glass moulds for matched moulding are made on similar lines to open moulds described but with the addition

Fig. 10.15. Matched dies for an outer panel in GRP. (Courtesy Pressed Steel Fisher BL Components Ltd)

of a male half, a plug. This is produced from the finished female half by lining it with a special wax sheet the thickness of which is equal to the wall thickness of the component to be produced. The male plug is then laid-up in a similar manner to the female until the required wall thickness is built up. Depending on its size and shape, the plug will require stiffening in the same way as the female mould. Again all ribbing and supporting structures should be built-in before the plug is removed from the wax. After removal the surface will require cleaning, polishing if any imperfect areas can be seen and finally waxing and treating with parting agent. Depending on the system of moulding to be adopted, for example, cold press, hot press, resin injection, etc. both halves of the mould will require to be suitably mounted.

Figure 10.15 shows the matched moulds for an outer panel mould mounted on baseplates and fitted with locating pillars and bushes.

Polyester/glass and epoxide moulds can incorporate means for heating one or both halves. Methods include (1) the use of hot water or steam channels in the form of copper pipes laid over the undersurface of the model and over laminated during construction and (2) low temperature electrical resistance wires embedded in the back-up laminate. When using some form of internal heating, such as described to speed cure, moulds can be made in epoxide glass using a high proportion of metal powder, usually aluminium, in the laminating resin to assist conduction.

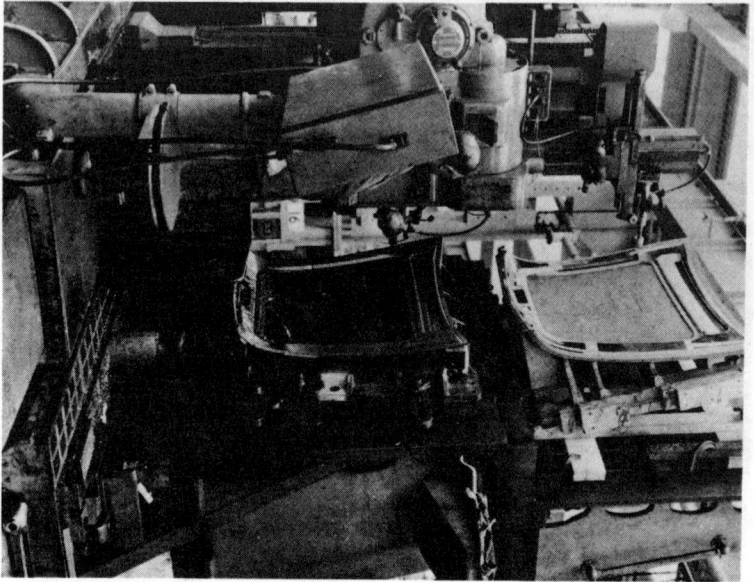

Fig. 10.16.   Die for an inner panel undergoing contour milling. (Courtesy Pressed Steel Fisher BL Components Ltd)

The construction of moulds for use in the vacuum and gravity systems of moulding follow the same sequence as described for matched GRP moulds with the difference that the male half is considerably thinner in order that it can flex somewhat during the application of the vacuum to consolidate the resin/glass laminate against the female mould surface.

10.9   MATCHED METAL MOULDS: HIGH PRESSURE SYSTEMS

Matched metal moulds used for SMC and DMC, and in long run situations for other press moulding systems, are to-day produced in specialist toolrooms using conventional metal die sinking equipment. It is outside the scope of this book to describe other than briefly the work involved in producing these tools. Should the reader wish to study the subject he should refer to the specialist literature.

It is usual for the body designer to produce the necessary component drawings and in some cases a master pattern as a starting point for the mould maker. The sequence of operations which follows is current practice in a well equipped shop. These are, briefly,

Fig. 10.17. Mould for an inner panel in the 'try-out' press. (Courtesy Pressed Steel Fisher BL Components Ltd)

Fig. 10.18. Matched, chromium plated, metal faced mould for an automobile console. (Courtesy J. Coudenhove Kunststoffe Maschinen GmbH)

(1) Shaping and marking out the billet which is normally of oil hardening steel or cast Meehanite — the latter offers the advantage that heating channels can be cored and considerably less machining is required.

(2) Milling of the moulding faces using a copy mill and working from a wooden, polyester or epoxide/glass master pattern, Figure 10.16.

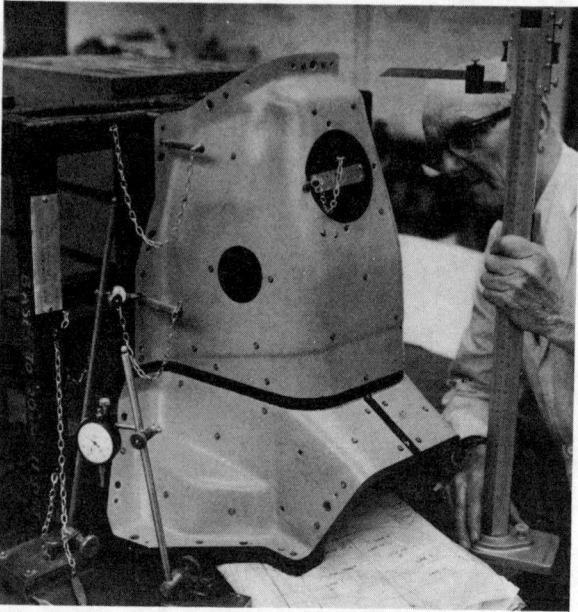

Fig. 10.19. Gearbox panel assembly checking fixtures for Standard Triumph by Simmons Patternmakers. Accuracy is to ±0.050 mm, using an Araldite system. (Courtesy CIBA (ARL) Ltd)

(3) After the die sinking operation, if required, side ram apertures and ejector apertures are machined and if not cast in, heater channels drilled.

(4) Polishing of the moulding faces is completed and, if required, the moulds are hardened.

(5) Moulding faces are finally polished, side ram and ejectors fitted together with their operating mechanism and the moulds screwed to baseplates or a die set, Figure 10.17, for mounting on the press.

10.10   EPOXIDE MOULDS AND CHECKING FIXTURES

Epoxide moulds can now be metal faced on the moulding surfaces using aluminium, copper or nickel spray techniques, and with the exception of aluminium sprayed moulds, can be chromium plated for optimum finish on the component. Figure 10.18 shows such a mould produced for an automobile console.

In addition to being used for long-life moulds, epoxide resins are also used extensively for the production of checking fixtures, being dimensionally stable and light to handle. A gearbox panel assembly fixture for Standard Triumph is shown in Figure 10.19.

11

Laminating tools and equipment

11.1 INTRODUCTION

Equipment can be divided into hand tools for hand lay-up of limited production mouldings and equipment for closed mould and high output production. The former are relatively simple and include items such as bruches, rollers, basic measuring equipment and hand tools. The latter category involves mixing, metering, spray-up, gel coat, resin injection equipment and SMC and dough moulding presses.

11.2 HAND TOOLS

During hand lay-up, random mat can be torn to the rough shape required or can be cut using a knife or a large pair of scissors. Glass fibre is highly abrasive and some operators prefer to cut the mat to shape with a replaceable blade or knife that can be easily sharpened.

The method used depends largely on the size and type of moulding being produced. Tearing the mat has the advantage that the join between abutting pieces is more easily and efficiently fused. If the sections are cut the edges must be teased out during lamination to avoid a weak area. If the operator's hands are completely clean of resin, rubber gloves or disposable polyethylene gloves can be used during the actual laminating operations.

Standard paint brushes are suitable for applying gel coats but brushes for working the resin into the mat should have reasonably short bristles as the action is that of stippling rather than brushing. Some operators prefer to use a lambs wool or mohair roller for gel coat application and a number also use one of these rollers for applying laminating resin and for effecting a degree of consolidation. All brushes and rollers must be thoroughly cleaned immediately after use. Various solvents are available including special polyester solutions, cellulose thinners and acetone. After cleaning in one of the above solvents a final wash in strong, warm detergent leaves brushes and lambs wool rollers in good condition.

For final impregnation and consolidation of both hand lay-up and spray-up laminates the most effective tool is a roller, of which there are a number of special types available. One type consists of stiff bristles or blunt ended spikes both of which have the effect of accelerating the wetting out process, consolidating the laminate and bringing any air bubbles to the surface. The most widely used roller, however, is the paddle type or washer roller which is build-up of metal or nylon washers spaced apart by smaller diameter washers threaded on a spindle. These vary both in diameter and in the number of washers, the choice of width depending on the size and nature of the mould. Plastics rollers can be cleaned in the same way as brushes but metal rollers can be heated to char the resin and then cleaned with a wire brush.

For very large work where the operator has to consolidate laminate some distance from the mould edge special 300 mm wide rollers of various types are available. These are fitted with a broom handle socket for long reach. A new type of spring roller is now available which consists of a deformable steel spring some 40 mm in diameter designed for use on internal or external curved surfaces. This type should not be heated for cleaning.

Trimming of freshly gelled mouldings can conveniently be carried out with a sharp knife or scissors. When the resin has cured, trimming requires the use of metal shears, a hacksaw or a Surform tool. Special files are available for working fully cured GRP and can be bought in flat, curved and half-round forms. Final trimming with an electric or pneumatic sanding machine is a rapid operation. Discs should be of the special open grit type or in a production shop, diamond edged or faced types as these last much longer than carborundum discs. It should be noted that in all dry-finishing operations a mask or better still, a respirator, should be worn as glass/resin dust must not be inhaled.

In some cases it is necessary to finish a mould or moulding using carborundum paper. This should be applied wet, commencing with a coarse grade such as 240 and finishing in the case of a mould surface with the finest of about 600 grit. Final surfacing of a mould, for example, should be carried out using a buffing machine in conjunction with a range of recommended 'soaps' available from suppliers.

11.3 SPRAY EQUIPMENT

Early equipment for spraying release agents, gel coats and laminating resin, although unsophisticated as compared with current equipment, was a great advance on the brush or lambswool roller methods particularly from the point of view of the speed of application that could be achieved. For this new application of the spray technique, equipment was based on conventional paint spray-gun principle. These guns operate on an external atomisation principle and thus have a high air consumption relative to the quantity of resin applied. A disadvantage to this type of gun is that it produces a considerable fog of atomised particles of resin which is wasteful, and, without adequate extraction, is a health hazard. A somewhat later type which came into use was the internal-mix gun. This reduced the 'backspray' hazard but by the nature of the principle required complete flushing immediately after use.

The latest type of gun operates on what can be termed the 'airless' principle, more specifically, by hydraulic atomisation. The three different principles now available to the laminator are:-

(1) The conventional, external air atomisation gun.

(2) The internal mix type in which atomisation takes place within the gun itself.

(3) The 'airless' or hydraulic atomisation gun.

Each type has its adherents but in general each is more suited to a particular type of job than the others.

11.3.1   External air atomisation type gun

In this method, air under pressure impinges on the resin stream to produce a fan shaped pattern as in a conventional paint spray-gun. Resin is fed to the gun by low-pressure pumps or alternatively from pressurised containers. Consumption of air by this type of gun is generally of the order of 0.56–0.70 m^3/min at the comparatively high line pressure of 3.5–4.9 kgf/cm^2. Application rates vary from about 400–900 cm^3/min and thus this type of equipment must be used in conjunction with a highly efficient system of fume extraction in order to comply with health regulations.

In addition to the disadvantages of overspray, many external air atomising guns can only be operated efficiently with the less viscous types of resin and have difficulty in handling filled and fire-retardant formulations. However, this type is usually less expensive than the 'airless' gun and is adequate for resin application to medium sized, generally flat moulds. As with the other types of gun mentioned the external air atomisation type can be used in conjunction with a roving chopper.

11.3.2 Internal-mix type gun

In this system both the resin and the air are mixed within the gun and at comparatively low pressure. Because the atomising air does not impinge on the resin stream as in the external atomising type, the internal mixing gun gives a round pattern rather than a fan pattern spray. Output is somewhat lower than the external type, being in the range 250–600 cm^3/min. These guns are thus mainly used for small intricate laminating work and as catalyst/resin mixing is particularly efficient, they are also used for the application of gel coats. As mentioned previously, their main disadvantage is that they must be installed in conjunction with a flushing system to avoid internal gelling after use.

11.3.3 Airless or hydraulic atomisation-type gun

With the introduction of filled, pigmented and fire-retardant resins equipment manufacturers appreciated the limitations of the external and internal mix systems and evolved a type of gun which uses a pump to generate hydraulic pressure. In this system, resin under considerable pressure is forced through the nozzle orifice to atomise without air impingement. The advantage of the system is that the finely atomised resin leaves the nozzle at a comparatively low speed and thus produces considerably less fog or overspray than the other types. Additionally, the fact that the resin is fed under pressure permits this type of gun to handle the more viscous, filled resins without difficulty and at substantially larger application rates. As the system involves external implosion of catalyst into the resin stream no flushing of the equipment is required after use. Outputs are considerably higher than are obtained from the foregoing types and typically range from 1200–1500 cm^3/min. These high application rates make the system ideal for the lay-up or large moulds and for use with automatic systems. In conjunction with an efficient roving chopper an airless gun can apply up to 5 kg/min of chopped glass. A final point in favour of the hydraulic atomisation system is that as it does not rely on the introduction of high pressure air to produce the spray, wetting out of the glass is efficient and the problems of excluding air during consolidation of the laminate are considerably reduced. The differences between these systems in operation and from the point of view of overspray and pollution are illustrated here.

Airless systems use hydraulic pressure alone to break the material down into droplets. Since there are no jets of compressed air to create turbulence in the spray pattern, there is little vapour fog, no catalyst

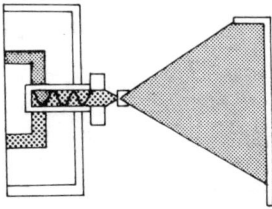

☐ Catalyst

▦ Resin

Fig. 11.1.   *Airless spray system*

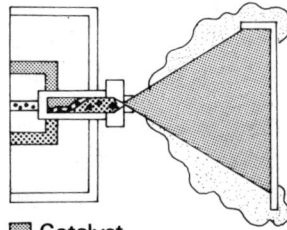

☐ Catalyst

▦ Resin

⊡ Air

Fig. 11.2.   *Air aspirated system*

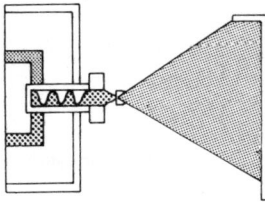

Fig. 11.3.   *Internal mix system*

☐ Catalyst

▦ Resin

loss and very low fumes. The system is shown diagrammatically in Figure 11.1. Air aspirated spray-up systems which use compressed air to atomise the material are characterised by the fact that the compressed air expands the spray pattern and carries finely atomised material into the surrounding air as a mist. In systems where catalyst is mixed with the atomising air, catalyst mist is carried into the atmosphere, Figure 11.2.

11.4   INTERNAL MIXING VERSUS EXTERNAL MIXING SYSTEMS

In the internal-mixing type gun in which a turbulent unit in the gun head performs the operation, no catalyst comes into contact with the air and thus no fumes are generated, Figure 11.3. Conversely, in an external mix system the catalyst must make contact with the air before it mixes with the resin. Because the catalyst is volatile,

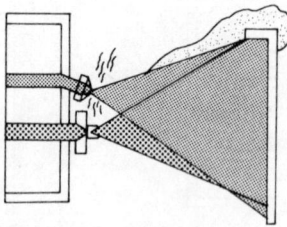

☐ Catalyst

☐ Resin

Fig. 11.4. External mix system

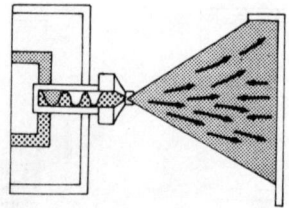

☐ Catalyst

☐ Resin

Fig. 11.5. Low pressure airless system

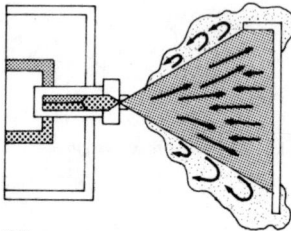

Fig. 11.6. Medium pressure airless system

☐ Catalyst

☐ Resin

quantities are lost as fumes, causing a fire and health hazard, Figure 11.4.

11.4.1 Low pressure airless versus medium pressure airless systems

Understandably, the lower the pressure in an airless system, the lower the generation of fumes and mist. The American Venus Hydraulic Injection System (HIS), shown diagrammatically in Figure 11.5, operates on a pressure of 14–21 kg/cm^2 which is said to result also in a minimum of air inclusion in the sprayed laminate leading to a reduction in the time required for roll out. On the other hand when medium hydraulic pressure is used to break the material down into droplets, the particles are propelled towards the mould at higher speeds and through friction with the air some mist and fume is generated, Figure 11.6.

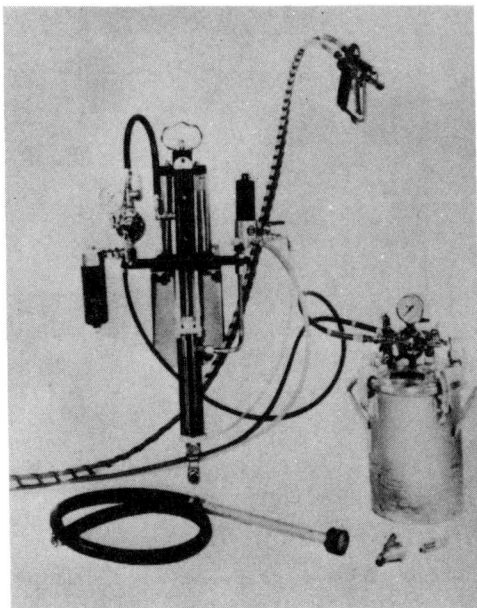

Fig. 11.7. Maverick gel coat unit. (Courtesy Prodef Engineers Ltd)

11.5 RELEASE AGENT AND GEL COAT SPRAY EQUIPMENT

Spray equipment for applying release agents is produced by a number of companies, the names of which appear in the Suppliers list. The Downland type R.A. gun is fitted with a small nozzle and needle for applying PVA and other low viscosity release agents. It enables a fine coating to be applied to the mould surface without any danger of runs to spoil the finish. The unit is fitted as standard with a 0.25 litre gravity feed cup and has the advantage that material can be left in the gun ready for use. A similar unit, type GC, is fitted with a large nozzle and a 0.5 litre Nylon cup for applying catalysed gel coats. A larger remote pressure container can replace the gravity cup on both units when it is necessary to spray larger quantities.

For continuous application of gel coats to larger moulds equipment such as the Downland type MGG, Prodef Maverick (Figure 11.7) or Coudenhove Polyspray M70 can be used. In the MGG gun resin is fed from pressure containers one of which contains accelerated resin, the other catalysed resin. The two mix internally prior to leaving the nozzle. An integral acetone flushing system ensures that

Fig. 11.8. Maverick airless resin/catalyst mix spray unit. (Courtesy Prodef Engineers Ltd)

both the mixing chamber and the spray head are thoroughly cleaned after use. This gun is available with hoses and a solvent tank for use with pressure containers and control panel on the Downland resin glass spray units described later, or as a complete unit comprising two 13.6 litre pressure containers and solvent tank mounted on a wheeled trolley. The Maverick airless external resin/catalyst mix spray gun (Figure 10.8) incorporates a 25:1 ratio airless pump providing material pressures up to 175 kg/cm². Material output is up to 0.945 litres/min. The unit can be used for single or multi-colour gel coat spraying, wetting out and spray-up with the addition of a glass fibre roving chopper. The high ratio of the pump enables an efficient spray pattern to be produced with most thixotropic gelcoats. The required catalyst atomisation pressure is by the use of a thinning agent, if necessary. Included in the unit are an air and water filter, a stainless steel catalyst container, resin filter with recirculation valve and 8 m of resin and catalyst hoses. Also produced by the company is a variant of the Maverick which incorporates a 30:1 ratio airless pump and capable of an output of 3.75 litres/min.

The Coudenhove Polyspray M70 operates on a combined resin catalyst pump system with a special high performance catalyst pump — a system which obviates the need to keep peroxide catalyst under pressure. The unit operates in conjunction with 30 litre resin and catalyst containers, gives a maximum spray width of 180 mm and a gel coat output of 600 to 1800 g/min. If the gel coat has to be brushed on rather than sprayed the Polyspray can be fitted with a brush. Catalysed resin is fed through the handle. A valve operates the feed and a flushing system is incorporated for cleaning. Brushes are clipped on to the handle and are readily interchanged.

In cases where frequent colour changes are required the Polyspray type M80 is recommended. This unit is provided with detachable pump cylinders. The cylinders, which are fitted with a suction hose and filter are readily connected to the piston of the pneumatic motor. The resin line from the cylinder to the gun is fitted with a rapid coupling so that it is a simple procedure to disconnect one cylinder from the piston and resin line from the gun and to attach a new cylinder for a different colour.

A Downland unit developed for gel coating where a large or continuous run of moulds is being handled is the type GP in which pre-accelerated resin and catalyst are both drawn from the resin manufacturer's containers. A built-in flushing system using a mixture of solvent and air purges the gun after use. Also to change colours it is only necessary to flush out the pump, hose and gun, and place the new cylinder of gel coat in the unit. The complete equipment is trolley mounted and has an output of 0.5–1.5 kg/min. The containers hold up to 50 kg of resin and up to 25 kg of catalyst. Air consumption is of the order of 0.34 m^3/min.

The Polyspray M60 airless gel coating equipment is designed for use where large moulds and few colour changes are required. The unit operates on the two component system now accepted as being the most reliable. The dangers of an error in setting the catalyst ratio are minimised since the gel coat is prepared in advance. In operation a specially developed high-pressure pump with one air cylinder and two pistons draws the two components from open containers. The addition of special inhibitors to the catalyst component provides for a pot life of about ten hours. The two components are pumped under pressure at a 1:1 ratio and thoroughly mixed in the static mixer of the spray gun. Output of the gun is determined by the size of the nozzle orifice. A solvent flushing system is part of the machine. Advantages of the system are a minimum of pollution from styrene in the spraying area, the capability of applying uniform and extremely thin gel coats with a minimum of air entrapment. Recommended gel coat thickness when using the system ranges from 0.4–0.5 mm.

11.6 AMERICAN GEL COAT SYSTEMS

Considerable development work has taken place in the U.S.A. on both gel coat and resin/glass spraying equipment. Features of the

Poly-Craft Systems, Model G-570 airless gel coat unit are a patented resin/catalyst system which is claimed to ensure uniform mixing regardless of material tip size, an instant-change 360° rotational nozzle which permits the operator to change from vertical to horizontal spraying, no requirement for dilutents and a universal type catalyst tip which handles all material fan widths. Because of the external mix system no equipment flushing is required. The equipment is available in completely portable form as Model G-565.

Plural Component Systems produce a Model 100 gel coat unit which with the addition of a roving chopper can also be used for spray-up applications. The unit is also available as a trolley mounted machine. Also portable is the company's Model 165 airless gel coat unit which operates at low pressure, mixing the two components of the resin 36 mm from the gun nozzles before the streams assume a fan pattern. Plural also produce a multi-colour unit designed to facilitate and speed colour changes by the flushing of 1 m of hose and the gun. The unit is available for from two to eight colours. Equipment by Venus Products Inc. operates on an hydraulic injection system (HIS), an internal mix, low pressure airless principle with surge control and precise metering characteristics. The HIS gel coating unit (Figure 11.9) provides for catalyst ratios of from 0.3–3.0% with an accuracy of 0.1%. Seven nozzle sizes vary from 0.5–1.0 mm giving outputs of from 1.2–3.5 kg/min. Air consumption is between 56 and 210 litres/min depending on the nozzle size, air pressure and volume of gel coat.

Equipment by Glas-Craft includes a system which utilises a special design of catalyst metering permitting catalyst injection of from 0.25–4.0% to be controlled easily. Resin and catalyst are mixed within the gun and sprayed airlessly through the nozzle. Roving is chopped by a high throughput chopper and fed externally into the resin and catalyst stream. During pauses in production the gun can be quickly and easily purged by pressing the purge buttons located on each side of the unit.

11.7   MEASURING AND METERING EQUIPMENT

While resin, catalyst and accelerator quantities can be measured using simple measuring cylinders it is a time consuming and often inaccurate operation and for serious production some form of more rapid and precise equipment is required. One such piece of equipment is the Polyspray M15. This unit is based on the Coudenhove double pump system which is described briefly in Section 11.7.1. In operation, pre-accelerated resin and catalyst are taken from unpressurised containers by the double pump and are fed through flexible hoses to

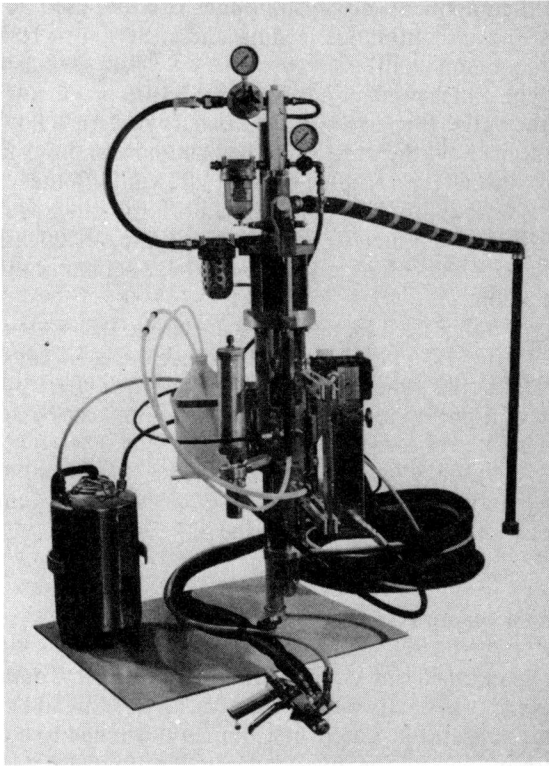

Fig. 11.9.   *Venus model HIS* 15-75-8000 *gel coater with temperature/viscosity control.*
(Courtesy Venus Products Inc.)

the static mixing head. The mixed resin is then poured into a suitable container and the system flushed with the integral solvent flushing system.

Metering equipment developed by the Research Institute for the Plastics Industry, Budapest, known as Ke Vi Ti, has multi-component dispensing units incorporating the facility for connecting any number of auxiliary pumps to the pneumatically operated resin pump. The additional pumps operate in synchronism without any mechanical connection. The required percentage ratios of the auxi-liary material in relation to the resin are adjustable on the auxiliary pumps thus the unit feeds each material with a high accuracy. The advantage of the pneumatic drive of the auxiliary is that an unlimited quantity of auxiliary material can be dispensed into the resin, without overpressure occurring, thus eliminating the risk of accidents. The Ke

Vi Ti multi-component dispensing unit can be used for both continuous and discontinuous feeding and metering of resins and their ancillary components.

The 'H' type unit is used mainly for hand lay-up processing. When operating the valve, the resin and its auxiliary materials flow into a suitable container. At the same time the pneumatic motor mixes the components inside the receptacle. After drawing off the required quantity the device stops automatically and the quantity drawn can be read off from a counter to within an accuracy of 0.1/litre. The required quantity of auxiliary materials can be re-set on the pumps or during operation.

The 'A' type unit is used for automatic metering. Its construction is identical with the 'H' type, the only difference being that instead of hand operation, the synchron valve works pneumatically and is provided with a pneumatic programme unit. This can be preset to between 1–2000 litres to within 0.2 litre accuracy.

In the U.K. Liquid Control Ltd produce the 'Twinflow' metering and mixing unit (Figure 11.10) designed for handling two-component materials of widely varying viscosities.

11.7.1 Double pump systems

Early mixing and metering equipment developed by J. Coudenhove operated on the twin pot (two batch) system, i.e. (1) resin plus catalyst, (2) resin plus accelerator. Later the Company switched to a catalyst injection (resin pump) for pre-accelerated resin with a catalyst pressure pot and flow meter. This system is the most widely used in the world today for almost all types of work. However, it can suffer from irregularities of catalyst content due to changes in catalyst viscosity, the disadvantages of catalyst under pressure and a lack of dependence between catalyst and resin flow. The latest design by Coudenhove is the combined resin/catalyst pump which ensures accurate metering of both resin and catalyst.

In operation pre-accelerated polyester resin with an almost unlimited pot life is pumped either from its original drum or from a central tank system. A flexible suction hose and filter eliminates the need for pump-lifting devices. Coupled to the resin pump is a stainless steel/Teflon catalyst slave pump which draws liquid MEKP catalyst from its original, non-pressurised container, Figure 11.11. The catalyst pump can be set to pump catalyst in a ratio to the resin of 1–4%. There is direct dependence between resin and catalyst pumping which is not affected by either resin or catalyst viscosity: one resin pump stroke is equal to one catalyst pump stroke.

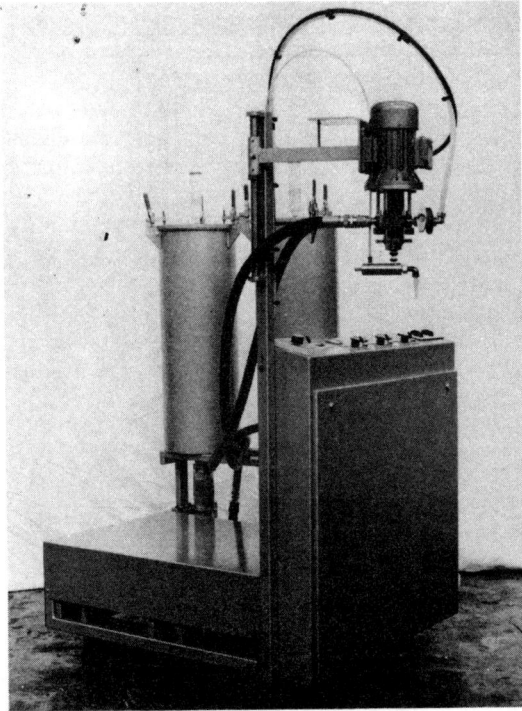

Fig. 11.10. 'Twinflow' metering and mixing unit. (Courtesy Liquid Control Ltd)

11.8 RESIN/GLASS SPRAY-UP EQUIPMENT

The use of a resin/catalyst spray gun in conjunction with a glass roving chopping unit, as described in Section 11.3, greatly speeds the laying-up particularly of large moulds in production runs. Additionally the automatic metering of resin, catalyst and chopped roving ensures a consistency difficult to obtain when handling the materials in a hand lay-up situation. In similarity with gel coat units, early resin/glass spray-up equipment was based on the conventional paint spray-gun principle with the addition of a simple and often inefficient roving chopping device. Today, all suppliers of equipment have available specially designed resin/glass depositing units which can lay down substantial quantities of material, with reliable and predictable results.

In the U.K., K & C Mouldings produce two basic models. The standard unit uses the separate catalysed and accelerated resin

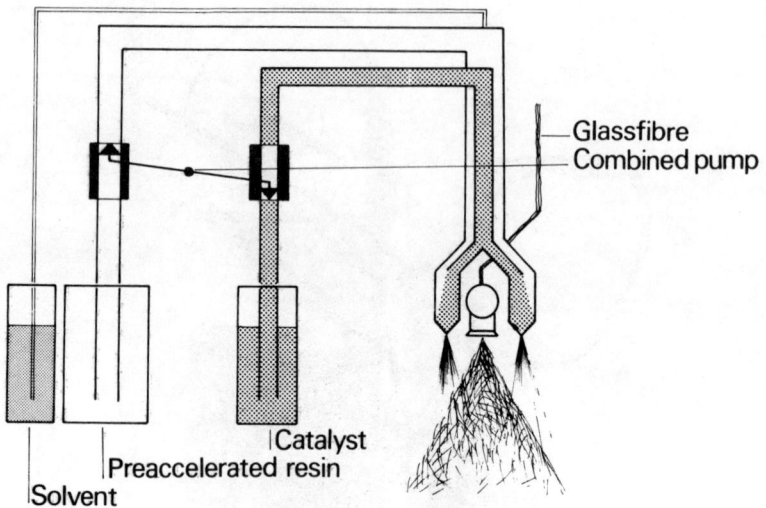

Glassfibre
Combined pump

Catalyst
Preaccelerated resin
Solvent

Fig. 11.11. *Coudenhove resin/catalyst metering system*

system, each component being fed via flexible hoses from individual pressure containers and remaining separate until combined with the chopped glass fibres sprayed from the top mounted depositor. The twin-head spray-gun has readily detachable heads for cleaning, all other parts being interchangeable. The glass fibre depositor is fed from a 'cheese' of roving and is driven by a variable speed pneumatic motor. It is mounted on the gun by an adjustable bracket in order that the angle of projection of the cut fibres may be altered. The length of the fibres can be controlled from 12.5–63 mm: the device is designed to handle 3 × 60 end roving. A rovings balance is supplied to permit the operator to check the weight of the rovings used.

Deposition of the resin and glass is accurately controlled by three air regulators and gauges on the control panel. Once set to suit the required resin viscosity and resin/glass ratio it will maintain an accurate output without further adjustment. The gun handles resins of any viscosity with or without fillers and resin/glass ratios of 2:1 are obtained at outputs of up to 5 kg/min. Roving length is adjustable as described. Units are supplied with 2 × 13.6 litre or 2 × 27 litre pressure vessels with separate inner resin containers for ease of mixing and cleaning. The vessels are trolley mounted complete with control panel, roving container and boom. Air consumption is 0.34 m³/min at 5.7–7.0 kg/cm².

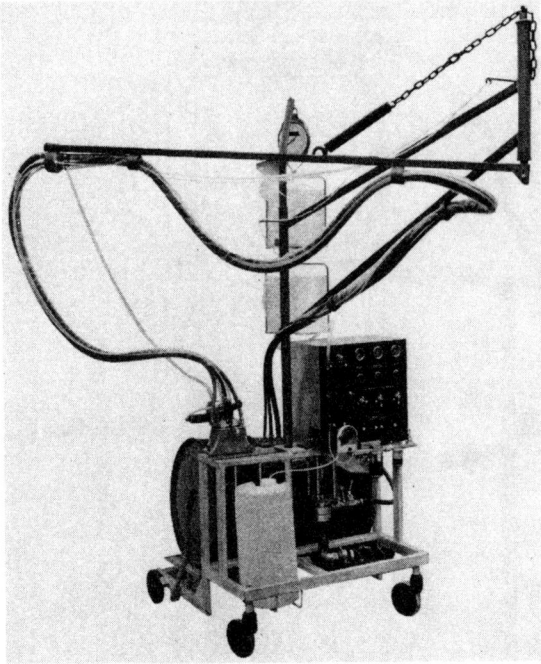

Fig. 11.12. *K & C Mouldings Type 'C' unit*

A higher output unit, Model C (Figure 11.12), incorporates a low pressure catalyst mixing system giving a resin output of 5 kg/min. Accelerated resin is pumped directly from a drum or bulk storage tank and is fed to a single head gun where it is catalysed automatically. The catalyst can also be drawn direct from the manufacturer's container and is fed into the resin through a hollow needle in the gun. In this system the resin and catalyst pumps are interconnected to give accurate metering. Catalyst ratios of $\frac{3}{4}\%$, $1\frac{1}{2}\%$ or 3% can be set whilst the unit is in operation. A feature of the unit is that the catalyst pump is valveless and is so designed that the catalyst is at no time under pressure. In addition the pump is reversible to return all catalyst in the system to the storage vessel after use. As the resin and catalyst mix externally there is no requirement for flushing the equipment. The glass fibre depositor accepts 1 or 2, 60-end roving, is operated by a high-speed pneumatic motor and can be adjusted to produce rovings of 12.7–60 m in length. Resin/glass output of the Model C is 75 kg/min.

The American company, Venus, also produce both reinforcement saturation (Figure 11.13) and spray-up units (Figure 11.14) both of

Fig. 11.13. *Venus Saturator unit.* (*Courtesy Venus Products Inc.*)

which operate on its hydraulic injection system. Standard equipment includes spray gun, catalyst pump capable of producing ratios of 0.5–3%, a resin pump, catalyst and resin accumulator, flow control and flushing system. Power head pressure ranges from 2.8–7.03 kg/cm^2, nozzles sizes from 1.28–2.2 mm giving outputs of from 4.5 to 10.5 kg/min.

The Venus spray-up equipment operates on a two-pot, external mix, airless system and is suitable for both small and medium work. It has low fume generation characteristics and is essentially simple to use.

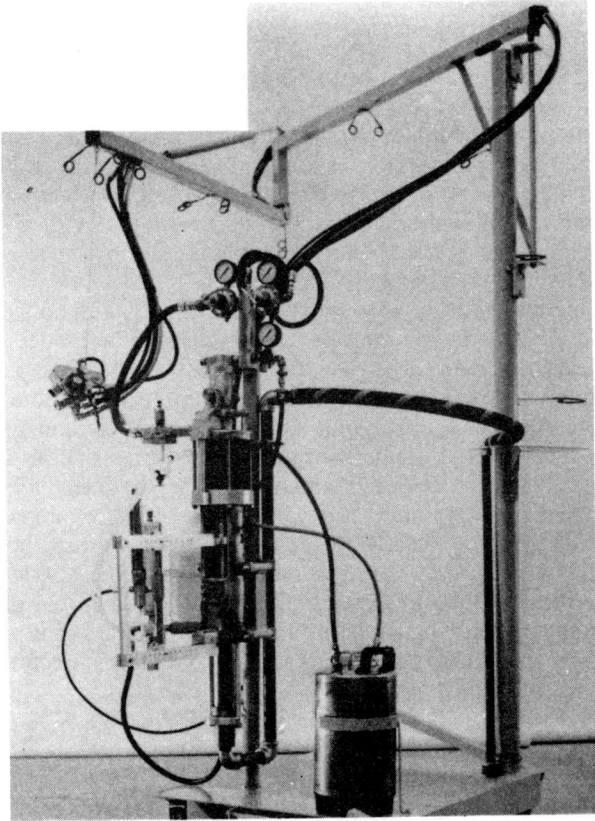

Fig. 11.14.   Venus Spray-up unit. (Courtesy Venus Products Inc.)

Resin/glass spray up equipment developed by the Austrian spe-
cialists, Coudenhove, includes an advanced unit the Polyspray M-40.
The system of resin/catalyst mixing is based on the company's twin
container coupled pump metering technique described in Section
11.7.1. Two spray heads are used to saturate the centrally ejected glass
fibres from both sides. Swivel mounted, the spray heads allow the
operator to select either a concentrated or fan-type spray pattern
according to requirements. A built-in solvent flushing system cleans
the spray heads. The roving cutter, driven by an 0.8 hp pneumatic
motor incorporates readily interchangeable anvil rollers and adjustment for fibre length to 17, 35 or 52 mm. Other features of the M-40
include a flow control device in the form of a photo-electric cell which
monitors the flow of catalyst after it has been metered by the pump.

Should flow be arrested or the container be emptied a warning signal is sounded. To monitor the quantity of resin and rovings used the machine embodies a pump stroke counter and scale which indicates rovings usage. The spray-up of large moulds is facilitated by the provision of a 4 m boom with a counter-balance and swivel joint which allows for a working radius, including bases, of up to 5 m. Maximum output with one roving is 4–4.5 kg/min, with two rovings up to 9 kg/min. A somewhat smaller, but similar, unit with an output of 3 kg/min is the model M4A which is suitable for use by less skilled operations.

In some cases smaller resin/glass depositor guns with only a single resin nozzle and roving chopper are required. The model Pl30 is developed specifically for narrow and intricate mouldings where only small outputs are required. Polyspray M40 units can be used with this gun. The American Poly Craft airless 'Avenger' system and air-atomised 'Fan-Jet' system incorporate a positive action catalyst system, as in the model G-570, which eliminates the need for flushing, permits instant start and stop with no waiting for pressures to stabilise and includes a safety exhaust to de-pressurise the entire system. The gun used is exceptionally light, and the nozzle or fan pattern rotates 360° and gives an output of 11.3 kg/min. The lightweight glass-roving cutter mounted on top of the gun is pivotted and has an adjustable thumb-screw to control glass direction.

11.9   CENTRAL RESIN SYSTEMS

For companies using large quantities of resin and a number of depositors a central resin system offers convenience and can result in considerable savings in both operator time and in materials. A central system such as that offered by Venus where a number of saturators and chopper guns are used, with various lengths of wall or ceiling mounted booms, is illustrated in Figure 11.15. Resin is stored in a large tank and circulated around the shop to the lay-up stations. In addition to obviating the need to use day tanks and more drums around the shop, further cost-savings can be made by bulk purchase of materials and in the elimination of contamination. Increased floor space is an additional benefit.

11.10   AUTOMATIC SPRAY-UP EQUIPMENT

Automation in the GRP industry is still in its infancy but with increasing demands on quality and the shortage of skilled labour

Bulk storage tank
Resin return
Circulating pump
Ceiling mounted boom
Solvent
Wall mounted boom

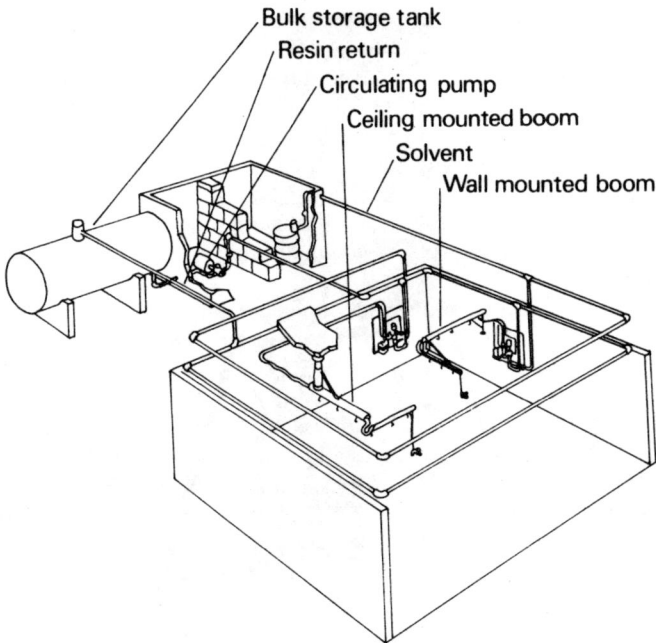

Fig. 11.15. Venus central control unit. (Courtesy Venus Products Inc.)

there is a rapidly growing need for automated equipment. Using manually operated spray guns, the quality and uniformity of the finished product depends to a great extent upon the skill and reliability of the operator. Automating and programming the operation guarantees uniform product quality. Currently automation of spray-up in the automobile body field appears limited but with developments in the process and in control methods it is an area that holds considerable potential for the processor able to handle production in sufficient quantities to make the system economic.

11.11   ROLLER IMPREGNATING EQUIPMENT

For all its advantages of speed of both resin and glass reinforcements a spray-up system cannot be used on all work and there are mouldings, generally large flat types, in which woven rovings are incorporated, where a pressure fed roller can be more efficient in wetting out than a resin spray gun followed by hand rolling. In the Downland system a resin/catalyst pump unit feeds and spreads

Fig. 11.16.   K & C Mouldings 'Autocat' gel coat roller unit

catalysed resin on to a wide mohair or lambswood roller, Figure 11.16. Flow of resin is controlled by the operator through a trigger on the roller unit. In this unit pre-accelerated resin and catalyst are drawn from the manufacturers containers and mixed in the head of the roller unit. A built-in flushing system using a mixture of solvent and air fully purges the unit after use. If require, a counter unit can be fitted in order that pre-determined quantities of resin may be dispensed to the roller.

The Coudenhove Polyspray M11 BX/1 is an extremely light dispensing unit for saturating mat and cloth which can also be fitted with a resin-fed roller or in the case of model M11 BX/2, with a resin-fed brush. In all cases where chopped strand mat or woven rovings or

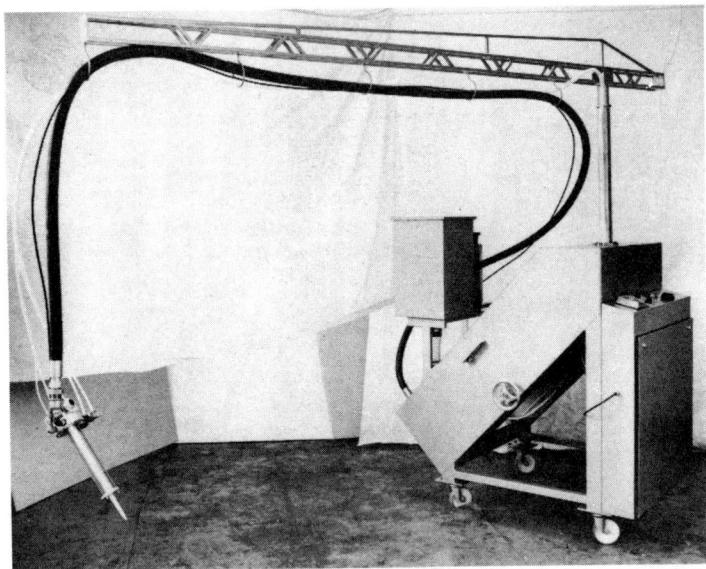

Fig. 11.17. *'Multiflow' resin injection unit by Liquid Control Ltd*

cloth are used as reinforcement these systems will show considerable savings over the conventional 'bucket and brush' methods.

11.12 RESIN INJECTION EQUIPMENT

The more recently developed resin injection equipment, as described in Sections 1.6.10, 4.10 and 4.10.2, closes the gap between manual methods of production such as hand lay-up and spray-up and mass production such as SMC and other press moulding methods. In addition, of course, it eliminates the environmental problems associated with 'free' styrene in the air.

Among leaders in the development of resin injection equipment in the U.K. are Liquid Control Ltd producers of the 'Turnflow' metering and mixing unit and of the 'Multiflow' injection equipment shown in Figure 11.17. This unit has been developed to meter and inject filled resins at highly accurate catalyst ratios at closely reproducible tolerances.

Other companies offering resin injection equipment in Europe are K & C Mouldings (England) Ltd., and J. Coudenhove GmbH, in Austria. The Downland equipment by K & C Mouldings (Figure 11.18) incorporates the company's standard twin-pump metering and

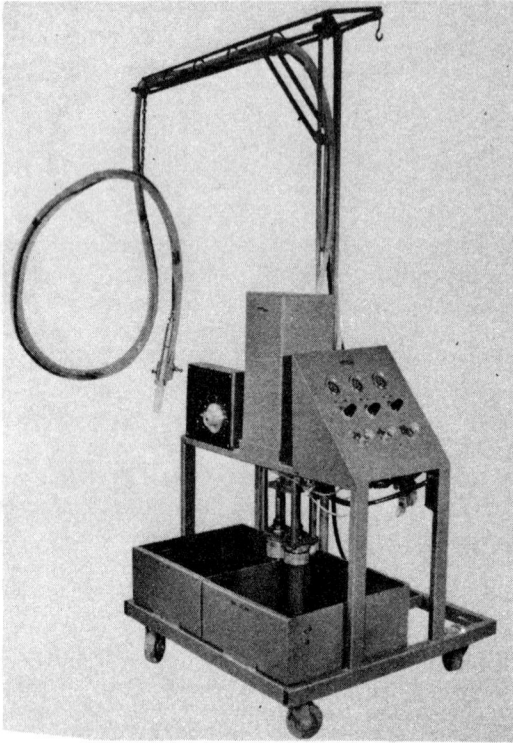

Fig. 11.18.   *Downland resin injection unit, by K & C Mouldings*

mixing machine in which the two resin pumps are linked to a single pneumatic motor. This enables equal volumes of each resin to reach the mixing head injection gun even if the viscosities of the two fluids vary. Speed of pumping and injection pressure are instantly adjustable. Output capacities of up to 7 kg/min can be obtained and mouldings with a high glass content are possible as there is ample pressure to force the resin through the glass fibre filled mould. The metering unit automatically measures the amount of resin injected thus there is no wastage beyond the mould faces through the need to overfill to ensure a completely injected mould. Because it is only necessary to flush the catalysed resin from the unit periodically, provision is made to carry out this operation while recirculating the accelerated resin, which may be left in the unit for any length of time without harm. If it is necessary to pigment the resin, the pigment need only be added to the catalysed resin as it will be completely mixed with the accelerated component in the injection gun, thus eliminating the need to flush out both resins when changing colour.

Other features of the Downland unit are the use of PTFE glands and pistons, an air and airline oiler to ensure correct lubrication of the pneumatic motor and mixer head, solvent flushing valves fed from a 13.6 litre solvent tank in the head of the gun to purge the mixing chamber after use and the facility for rapid detachment of the mixer unit from the valve block without disturbing the resin or solvent hoses. The weight of the injection gun is 3.5 kg with 6 m length of hose on the transportable unit.

In the Coudenhove Polinjector M16, pre-accelerated resin and liquid catalyst are drawn from their original containers as in the system described in Section 11.7.1. The two components are statically mixed and are injected through a teflon nozzle. A solvent flushing system is incorporated together with a photocell and acoustic warning signal as used in the company's spray-up equipment. Air consumption is of the order of 150 litres/min at a pressure of 6 kg/cm^2, catalyst ratio can vary from 1–4% depending on requirements. Resin output is dependent upon the size and design of the mould.

The Venus 'Hydrojector' injection equipment incorporates the HIS resin/catalyst feeding and metering system previously described in Section 11.8 and is capable of outputs of from 0.45–4.5 kg of resin per minute. This output is normally adequate for mouldings up to 3.5 m^2 of surface area. For even larger parts a high volume unit is available with 2–3 times greater capacity. The equipment, Figure 11.19, includes a 4:1 resin pump with a 76 mm air motor giving sufficient power for pumping most resins. A high volume pump is optional. There are accumulators for resin and catalyst, an integral solvent flushing system with an 11.5 litre tank and air gauges calibrated in both kg/cm^2 and lb/in^2. The catalyst metering pump controls catalyst/resin ratio to within ±0.1% and adjusts from 0.5–3% of catalyst. The catalyst inlet hose is fitted with a special wand which is placed in the drum to minimise waste and fumes. Injection pressures can be adjusted over the ranges 0.35–1.4 kg/cm^2 and 2.1–4.2 kg/cm^3 depending on mould requirements.

11.13 FILLER SPRAYING EQUIPMENT

The addition of a suitable filler in the correct ratio to the resins can improve the compressive strength of the moulding, give it increased bulk and through colour. The addition of a filler will, however, increase the viscosity of the resin and after incorporation will tend to settle and form a sludge if not continually stirred. In hand lay-up operations this characteristic is no great problem as the quantities of resin mixed at any one time are necessarily modest due to limitations

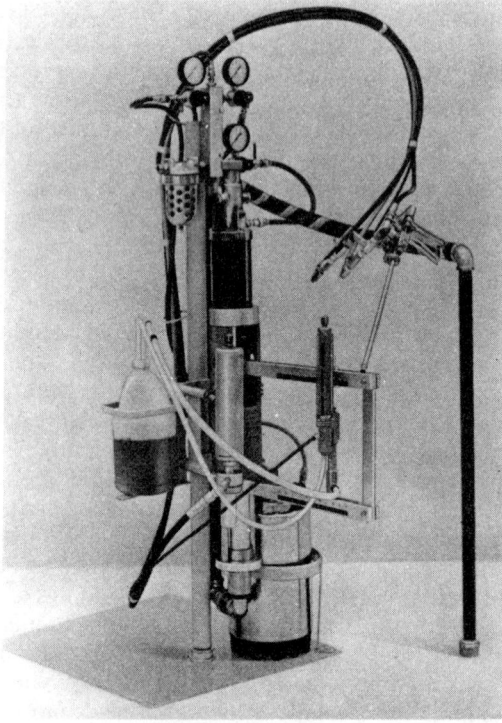

Fig. 11.19.   *Venus 'Hydrojector' resin injection unit*

placed on the quantity by gel time. In spray-up work where large
quantities of resin are being used the hand or mechanical mixing of
fillers is less practical both from the point of view of the formation of
sludge in the resin container and the added problems of spraying the
more viscous material.

Recently these difficulties have been overcome by Venus Products
Inc., in the U.S.A. by the introduction of its 'Dri-Adder' unit which is
designed to work in conjunction with the company's resin depositor.
It operates by feeding the powdered filler into a high volume
airstream which carries the powder through a separate hose to the
spray gun. The filler is then propelled into the resin/glass spray so that
it is deposited evenly throughout the laminate. The air stream is taken
direct from the main air supply and is controlled by a regulator. A
gauge is fitted to permit the operator to ensure that the pressure in the
hose remains constant.

The filler material is stored in a hopper fitted with a shutter which closes it off from the airstream. The shutter is operated by an air cylinder linked pneumatically to the roving chopper unit on the spray gun. In this way the trigger on the gun controls the feed of resin, glass and filler. The shutter aperature is adjustable so that the quantity of filler gravity fed into the airstream can be controlled by the operator. The filler unit can be rapidly fitted to the spray gun and brings the use of fillers within the scope of spray-up equipment users.

The 'Polyfiller' by Coudenhove is designed for attachment to the M40 spray-up equipment described in Section 11.8. In addition to adding filler to the sprayed resin and chopped glass rovings it can be used to produce sandwich construction laminates. Used in this manner the sandwich is built up by spraying inner and outer skins of polyester and chopped glass reinforcement over an inner core of polystyrene foam and filler only. The unit incorporates a trolley mounted hopper fitted with a pneumatically operated screw conveyor which transports the dry aggregate from the hopper to a plenum chamber and thence to the spray gun. An additional vibrator prevents clogging. This method of producing a light, high stiffness moulding is still in its infancy but obviously holds considerable potential from the point of view of speed and simplicity.

11.14   SAFETY EQUIPMENT

In the early days of GRP moulding few companies appreciated the need for safety equipment for operators and few operators realised the dangers that could arise from inhaling resin overspray and/or finely divided glass fibre carried in the workshop air. Pollution of the shop atmosphere can also arise from free styrene which is given off as a moulding cures.

Since the 1974 Health and Safety at Work Act and its provisions for regulations on air pollution GRP processors have been alerted to the dangers and to the problems of eliminating them. Apart from glass dust, the two culprits are styrene from the resin and the catalyst. So far the recommendations regarding styrene, the main offender, are that the threshold should be 100 parts in one million while regarding MEK peroxide catalyst the 1976 regulations prescribe a threshold limit volume of 0.2 parts per million.

As described previously air pollution is closely related to droplet size and impingement speed and in this respect the development in spray technology has gone a long way to reducing the amount of resin fog in the air. However, most extraction systems in use in GRP moulding shops operate on similar principles to those designed for air extraction in cellulose paint shops in that the extractor fans are located well above the work area. Styrene is heavier than air and thus these types of extraction system tend only to remove the heated air from the area leaving a styrene concentration lower down. This is both expensive from the shop heating point of view and inefficient from a health standard. Thus in a GRP processing shop the most efficient type of extraction equipment is one which removes air at a point about 1 m above ground level and has an input at a high level at the opposite side of the work area.

11.14.1 ANCILLARY SAFETY EQUIPMENT

The use of acetone and other highly inflammable solvents in the GRP industry present a very real fire hazard and carelessness in their handling has without doubt been the cause of many serious moulding shop fires. The following range of special containers by K & C Mouldings have been designed to meet the requirements of the Factories Acts relating to highly inflammable liquids. A 13.6 litre treadle operated dip tank, centrally sited in the shop serves several operators for cleaning brushes and rollers. The hinged lid is self-closing to minimise evaporation and in the event of ignition will close immediately the operator releases the pedal. Smaller safety dip cans also with a drop lid are available as are liquid tight flammable waste bins with self closing lids, special safety storage cans in heavy gauge tinplate with flame arrester pourers and a range of specially designed drum taps and pourers.

11.14.2 Protective, disposable clothing

Purpose designed personnel clothing for use in the moulding shop are indispensible for both the health and safety of operators. Full length, zip fronted overalls, gauntlets, overboots and elasticized face aperture hoods made from lightweight woven plastics material protect against resins, glass fibres and dust from trimming operations. Also available are aprons, rubber and disposable 140 gauge polyethylene gloves. Lightweight dust masks should always be worn when dry trimming is undertaken. These and a more sophisticated respirator effective against styrene and acetone is the Martindale-Fumex respirator with renewable filter pad.

Other essentials in the GRP shop are the special vanishing type multi-purpose creams which applied to the hands and face form a protective film against polyester and epoxy resins, urea formaldehyde, phenolic resins and irritant dusts and vapours. These hand cleansers are specially formulated for use by workers in moulding shops for removal of synthetic resins from their hands.

11.15   FINISHING EQUIPMENT

As mentioned briefly in Section 4.2.3, there are a number of methods of trimming and finishing hand and spray-up mouldings. 'Green' mouldings which have gelled but not fully cured can best be trimmed roughly with a sharp knife before removal from the mould. Later, when the resin has cured, shears or one of the many electric or pneumatic hand tools available give quicker and more accurate results. Cured GRP mouldings are highly abrasive and thus equipment such as saws must be of high quality and capable of being sharpened at regular intervals. Electric and pneumatic jig saws are ideal for initial trimming of cured mouldings. Electric saws, available in 280 W, 220–250 V models, are excellent both for pattern making, cutting metal, wood and GRP. They are available with a variety of blades. For cutting GRP the recommended Rynson blade type is T118G or type T150 medium or T150 coarse Riff blades for long life under difficult conditions. The machine weighs 2.26 kg and operates at 3000 strokes/min. The Lesto pneumatic jig saw is similar to the electric version but smaller and lighter, at 2.5 kg, and has the advantage of being safe to use under wet conditions where water is used to prevent dust problems. Air consumption is 0.24 m³/min at a pressure of 3.9 kg/cm² or 0.35 m³/min at 6 kg/cm². Stroke speeds are 1300 and 1800/min respectively. The 'Cengarette' glass fibre saw and air operated trimmer is also a lightweight, versatile tool for cutting GRP laminate. Using standard hacksaw blades and cutting at 10,000 strokes/min, the tool operates on 5.7–7 kg/cm² pressure at a consumption of 0.16 m³/min.

Two types of rotary laminate cutting and trimming tools are produced by the Austrian company, Coudenhove. The 'Polycutter Air 1', is a highly efficient, heavy duty pneumatic cutter incorporating a diamond tipped rotary steel blade of 100 mm diameter capable of cutting laminate up to 20 mm in thickness. The unit is fitted with a dust extractor shroud, operates at 1200 rev/min and weights 1.7 kg. The model 'Air 2' is smaller, designed for cutting laminate up to 6 mm in thickness and is fitted with a diamond tipped disc 50 mm or 63 mm in diameter. Operating speed is 4000 rev/min. The company also produce a trolley mounted, electrically driven cutter in which the cutting disc is mounted at right angles on a flexible shaft for ease of operation. Motor power is 6000 W and laminate cutting capacity 12 mm.

Coudenhove have developed special power tools for cutting glass fibre mat and cloth. Type B1 can handle cuts of up to 12 layers of 450 g mat and is widely used for preparing pre-cut sections of mat and cloth for the injection process and for cold press moulding. Type B2, also pneumatically operated, uses a smaller cutting blade and has a capacity for cutting up to 5 layers of 450 g mat. It is used mainly for the more difficult cutting jobs by virtue of its smaller diameter blade.

11.16 PRESSES

11.16.1 Low pressure presses

The pressure for moulding glass reinforced polyester, other than SMC and DMC, are comparatively low and for this reason quite large mouldings can be produced on low tonnage machines providing that the platen area and opening (daylight) is sufficient. However, the majority of presses used in the plastics industry are designed to provide pressures on the mould of the order of 1500–300 kg/cm^2 and thus have small platens. Presses designed for moulding by the preform and other methods have large platens (120 × 120 cm), large opening strokes and incorporate hydraulic systems and controls which permit the timing and speed of opening and closing to be programmed.

11.16.2 High pressure presses

Presses for moulding SMC and DMC require considerable higher pressure ratings particularly where deep draw mouldings are produced. In practice optimum moulding pressure is established after consideration of such variables as viscosity of the SMC resin, the complexity of the component and to a certain extent the quality of the moulded surfaces. Owens-Corning recommend that viscosities of low shrink, low profile SMC must be relatively high to ensure good flow characteristics particularly when large components are moulded. When large, deep draw components are involved pressures of the order 140 kg/cm^2 are required. Large, generally flat components require somewhat lower pressures to provide satisfactory flow and fill-out. The closing speed of these presses are dependent on mould temperature and the gel time of the SMC. As would be anticipated, higher moulding temperatures and rapid cure SMC formulations require faster closing speeds and rapid build-up of moulding pressure

Fig. 11.20. *Compression presses for large volume production of SMC components.* (*Courtesy Owens-Corning Fiberglas Corporation*)

on the material. Exact closing speeds, normally less than 10 s and in some cases less than 1 s are used to prevent pre-gelling of the resin which can occur if closing is too slow.

Presses for SMC, such as shown in Figure 11.20, must also incorporate hydraulic circuit controls to provide for rapid initial approach followed by final slow closing. This latter movement permits the material to flow in the mould before full pressure is attained and cure commences.

11.16.3 Press design

In view of the high pressures and close mould tolerances involved in SMC moulding the design of the press must be such that the alignment system between platens ensures rigidity and accurate and consistent location in order to prevent damage to mould shear edges. If accurate alignment of the moving and stationary platens by the pillars can be ensured even under side loading produced on mould closing the need for guiding shoes in the mould to align both upper and lower halves can be eliminated.

Another vital factor in SMC press design is parallelism of the platens if uniformity of thickness in the mouldings is to be ensured in high pressure moulding cycles. Special purpose presses are now designed with provision for rigid guiding as the point of material flow in the mould is approached and with retractable stops which force the platens into parallelism as full moulding tonnage is applied. The stops are then retracted to allow the material to cure under the full pressure of the press.

REFERENCES

1. *Crystic Monograph*, No 2, Scott Bader Ltd., Wollaston Wellingborough, Northants (1977)
2. *Beetle Polyester Resins Handbook*, CG100/0176/M/5M, British Industrial Plastics, Chemicals Division, P.O. Box 6, Popes Lane, Oldbury, Warley, West Midlands (1978)
3. *FRP in Automotive Construction*, GFK Conference, Dusseldorf. GFK im Fahrzeugbau, 115 Grafenburger Allee, 4000 Dusseldorf, West Germany (1978)
4. 'The future — an all plastics car?', *Plastics Design Forum*, Vol. 2, No. 4, Industry Media Inc., Denver, Colorado, USA (1977)
5. 'The plastics Ford in your future', *Plastics Design Forum*, Vol. 3, No. 1, Industry Media Inc., Denver, Colorado, USA

BIBLIOGRAPHY

'Airless or conventional spray-up — which is best?', *Reinforced Plastics*, June (1977), p. 170

Fibreglas Plastics Design Guide, Owens Corning Fiberglas, No. 5-PL-6960, (1975)

Giles, J. G., *Body Construction and Design*, Vol. 6, Pentech Press (1971)

Hartley, J., 'Painting bodies before they exist', *The Engineer*, March, 10 (1975)

Johnson, A. F., *Engineering Design Properties of GRP*, British Plastics Federation, HMSO, No 215/1

'Mechanisation — the only way to go', *Reinforced Plastics*, November (1978) p. 363

Montella, R., 'Designing for rigidized acrylic', *32nd Annual Technical Conference, Reinforced Plastics/Composites Institute*, Soc. Plastics Industry, Inc., USA (1977)

Plastics Machinery and Equipment Magazine (monthly publication), Industry Media Inc., Denver, Colorado, USA

Poston, I. E., 'Thermoset and reinforced plastics in the automotive industry', *National Plastics Conference*, Soc. Plastics Industry, Inc., USA, December (1976)

Reinforced Plastics: Guide to Hand Spray-up, Owens Corning Fiberglas, No. 5-PL-3101-E (1977)

Russell, L. M., 'FRP cold moulded with acrylic shells for intermediate runs' *32nd Annual Technical Conference, Reinforced Plastics/Composites Institute*, Soc. Plastics Institute Industry, Inc., USA (1977)

Smith, W. S., 'Plastics bodywork for truck cabs', *Europlastics*, April (1973)

Turner, S., *Mechanical Testing of Plastics*, Imperial Chemical Industries, Plastics Division (1973)

'Vacuum injection process for large GRP mouldings', *Reinforced Plastics*, October (1977), p. 321

Vogelei, R. A., 'Corvette plastic progress enters third decade', *SAE International Automotive Engineering Meeting*, Toronto, Ontario (1974)

APPENDIX. GRP BODYWORK SUPPLIERS

Codings: Abrasive papers, bands and holders, 1; Brushes for hand lay-up, 2; Catalyst and accelerator dispensing equipment, 3; Diamond tipped trimming tools, 4; Embedded heating elements, 5; Etch primers, 6; Expanded core material, 7; Fillers, 8; Fire retardant additives, 9; Foam core materials, 10; Gel coat spray equipment, 11; Glass fibre reinforcement, 12; Metal powder fillers, 13; Mould polishes, 14; Mould release agents, 15; Nickel moulds, 16; Pattern and mould makers, 17; Polyester film, 18; Polyester resins, 19; Resin injection equipment, 20; Rollers, 21; Spheres (hollow) silica, 22; Spray-up equipment, 23; Talc, 24; Thixotropic additives, 25.

Suppliers	Product
Alger-Sanders Ltd., 5 Arden Grove, Harpenden, Herts., England	1, 4, 21
Armoform Marketing Ltd., 38 North Bar Within, Beverley, Yorks., England	22
BA Chemicals Ltd., Chalfont Park, Gerrards Cross, Bucks., England	8, 9
Baxenden Chemical Company Ltd., Paragon Works, Accrington, Lancs., England	7
Bewax Products Ltd., 24 Sudley Road, Bognor Regis, Sussex, England	15
BMS Plastics Ltd., Higgins Lane, Burscough, Lancs., England	2, 8, 9, 11, 14, 15, 21, 23, 24, 25
BP Chemicals Ltd., Belgrave House, Buckingham Palace Road, London, England	19, 25
British Industrial Plastics Ltd., PO Box 6, Popes Lane, Oldbury, West Midlands, England	19, 25
J. Coudenhove, GmbH, Probst-Peitlstrasse 60, A-2103 Langenzersdorf/Vienna, Austria	3, 11, 20, 23
C.T. (London) Ltd., Venus Division, 3 Hobart Place, London S.W.1, England	11, 20, 21, 23
Dufaylite Developments Ltd., Cromwell Road, St. Neots, Hunts., England	7
EHE Ltd., Harleston, Norfolk, England	16
Ferro (Great Britain) Ltd., Wombourne, Wolverhampton, England	14, 15
Fiberglass Ltd., St. Helens, Lancs., England	12
Fi-Glass Ltd., Station Road, Edenbridge, Kent, England	17
Fillite (Runcorn) Ltd., 12 Astmoor Industrial Estate, Runcorn, Cheshire, England	22
R. J. Frankiss (Patterns) Ltd., Market Road, Richmond, Surrey, England	17
J. J. Harvey (Manchester) Ltd., Oldhem Street, Denton, Manchester, England	17
ICI Ltd., Ship Canal House, King Street, Manchester, England	7

ICI Ltd., Plastics Division, Welwyn Garden City, Herts., England	18
Impag (GB) Ltd., Lyon Industrial Estate, Kersley, Bolton, Lancs., England	10
JCB Plastics Ltd., Mill Street East, Dewsbury, Yorks., England	22
K & C Mouldings, Spa House, Shelfanger, Diss, Norfolk, England (Downland Equipment)	1, 2, 3, 4, 5, 7, 8, 10, 11, 13, 14, 18, 23, 24, 25
Kendia Ltd., Union Mill, Cranbrook, Kent, England	4
Liquid Control Ltd., 25 Harcourt Street, Kettering, Northants., England	20
Melbourne Chemicals Ltd., Plastichem House, Esher, Surrey, England	8
J. Marlow, Deddington Mill, Deddington, Oxford, England	21
Mitchigan Chemical Corp., c/o William Blythe & Co. Ltd., Church, Accrington, Lancs., England	9
Newgate Simms Ltd., PO Box 32, Chester, England	3, 11, 23
Norwegian Talc (UK) Ltd., 251 Derby House, Exchange Buildings, Liverpool, England	8
Novadel Ltd., St. Ann's Crescent, London SW18, England	3
Owens Corning Fiberglas, Fiberglas Tower, Toledo, Ohio, USA	12
Plastichem Ltd., Plastichem House, Esher, Surrey, England	9, 24, 25
Plural Components System Inc., Santa Ana, California, USA	23
Polycraft Systems Inc., Sun Valley, California, USA	23
Prodef Engineers Ltd., 1163 Bristol Road South, Northfields, Birmingham, England	23
Scott Bader Ltd., Wollaston, Wellingborough, Northants., England	19, 25
Shell Chemicals Ltd., Manchester Road, Manchester, England	25
Shell Chemicals Ltd., 1 Northumberland Avenue, London, England	19
Shelter Islands Co. Ltd., 42 Redland Grove, Carlton, Nottingham, England	11
Strand Glass Co. Ltd., Brentway Trading Estate, Brentford, Middlesex, England	1, 2, 3, 6, 7, 8, 10, 13, 14, 15, 18, 21, 23, 24, 25
John E. Sturge Ltd., Wheeleys Road, Birmingham, England	8
Synthetic Resins Ltd., Edwards Lane, Speke, Liverpool, England	19
Treplade Ltd., 137 High Street, Epsom, Surrey, England	14, 15
Trylon Ltd., Wollaston, Wellingborough, Northants., England	15
Warwick Chemicals, 54 Willow Lane, Mitcham, Surrey, England	25

INDEX